# Mathematical and Statistical Methods for Actuarial Sciences and Finance

Cira Perna • Marilena Sibillo
Editors

# Mathematical and Statistical Methods for Actuarial Sciences and Finance

Springer

*Editors*
Cira Perna
Department of Economics and Statistics
University of Salerno
Fisciano, Salerno, Italy

Marilena Sibillo
Department of Economics and Statistics
University of Salerno
Fisciano, Salerno, Italy

ISBN 978-3-319-05013-3 ISBN 978-3-319-05014-0 (eBook)
DOI 10.1007/978-3-319-05014-0
Springer Cham Heidelberg New York Dordrecht London

Library of Congress Control Number: 2014933795



Printed on acid-free paper

Springer is part of Springer Science+Business Media (www.springer.com)

# Preface

This volume[1] aims to collect new ideas presented in form of 4-pages papers dedicated to mathematical and statistical methods in actuarial sciences and finance. The cooperation between mathematicians and statisticians working in insurance and finance is a very fruitful field and provides interesting scientific products in theoretical models and practical applications, as well as in the scientific discussion of problems of national and international interest.

From the theoretical and applicative point of view, the topics covered in the book are: actuarial models; alternative testing approaches; behavioural finance; clustering techniques; coherent and no-coherent risk measures; credit-scoring approaches; data envelopment analysis; dynamic stochastic programming; financial contagion models; financial ratios; intelligent financial trading systems; mixture normality approaches; Monte Carlo-based methodologies; multi-criteria methods; nonlinear parameter estimation techniques; nonlinear threshold models; particle swarm optimization; performance measures; portfolio optimization; pricing methods for structured and non-structured derivatives; risk management; skewed distribution analysis; solvency analysis; stochastic actuarial valuation methods; variable selection models; time series analysis tools.

In the light of the successful cooperation between the above two quantitative approaches, the Editors of the volume organize the biennial conference on Mathematical and Statistical Methods for Actuarial Sciences and Finance (MAF), born at the University of Salerno in 2004 and just arrived at its 6th edition this year.

Salerno
April 2014

Cira Perna and Marilena Sibillo

[1]Published with the contribution of Dipartimento di Scienze Economiche e Statistiche, Università degli Studi di Salerno.

# Contents

# Can Personal Dependency Paths Help to Estimate Life Expectancy Free of Dependency?

**Irene Albarrán, Pablo Alonso, Ana Arribas-Gil, and Aurea Grané**

**Abstract** The aging of population is perhaps the most important problem that developed countries must face in the near future. In fact, one of the eight tackling societal challenges of the European program Horizon 2020 is concerned with it. Dependency can be seen as a consequence of the process of gradual aging. Therefore, its prevalence on the population, its intensity and evolution over the course of a person's life have relevant economic, political and social implications. From data base EDAD 2008 the authors constructed a pseudo panel that registers personal evolution of the dependency scale according to the Spanish legislation and obtained individual dependency curves. In this work, our aim is to estimate life expectancy free of dependency using categorical data and the functional information contained in these trajectories.

**Keywords** Dependency · Functional data · Life expectancy

## 1 Introduction

When talking about *dependency* two fundamental aspects must be considered. Firstly, the definition itself. Resolution R(98) of the Council of Europe defines dependency as "such state in which people, whom for reason connected to the lack or loss of physical, mental or intellectual autonomy, require assistance and/or extensive help in order to carry out common everyday actions". Secondly, the *assessment*

I. Albarrán · A. Arribas-Gil · A. Grané (✉)
Universidad Carlos III de Madrid, Getafe, Spain
e-mail: aurea.grane@uc3m.es

I. Albarrán
e-mail: irene.albarran@uc3m.es

A. Arribas-Gil
e-mail: ana.arribas@uc3m.es

P. Alonso
Universidad de Alcalá, Alcalá de Henares, Spain
e-mail: pablo.alonsog@uah.es

C. Perna, M. Sibillo (eds.), *Mathematical and Statistical Methods for Actuarial Sciences and Finance*, DOI 10.1007/978-3-319-05014-0_1,

*of dependency*, which is usually solved using specific dependency rating scales that take into account the disabilities suffered by the person jointly with their intensity. In Spain, the definition of dependency is that included in article 2 of Act 39/2006, of 14th December, on the Promotion, Personal Autonomy and care for Dependent persons and its evaluation is ruled by the Royal Decree 504/2007. The Spanish dependency rating scale goes from 0 to 100 points and it is categorized in four degrees (non dependant, I-moderate, II-severe, III-major). To acknowledge the entitlement to the benefits of the System a person must reach at least the moderate degree. According to the dependency rating scale value reached by an individual, the Spanish legislation establishes a minimum level of protection, which is defined and financially guaranteed by the General State Administration.

EDAD 2008 is the most recent Survey about Disabilities, Personal Autonomy and Dependency Situations conducted in Spain by the Spanish National Institute of Statistics and it is the first Spanish survey that uses the internationally accepted measures established by the International Classification of Functioning, Disability and Health. Following the World Health Organization recommendations, the survey is based on the concept of self perceived disability and, despite its drawbacks, the main advantage is that it focuses the attention on the daily activities of the individuals and the problems they may have while doing them, with no consideration of medical matters. According to EDAD 2008, there are more than 4.1 million Spanish people suffering at least one kind of disability. Although the global prevalence rate is situated between 8.2 %–8.6 % with a 95 % of confidence, in the case of people living at home, this rate is lower than that for people living in institutions (8.4 % and 17.7 %, respectively). Disability is related to two main factors: gender and age; until 45 years old, the male prevalence is statistically significant greater than the female one. After that age, the relative incidence is greater for women. In general terms, more than 57 % people with this problem are at least 65 years old, being most of them women.

Using a pseudo panel constructed from EDAD 2008, the main aim of this work is to estimate life expectancy free of dependency (LEFD), that is, the expected number of years that a person can live free of this contingency.

## 2 Methodology

In [1] a pseudo panel that registers personal evolution of the dependency scale according to the Spanish legislation was constructed. Using this pseudo panel these authors obtained functional profiles from the dependent Spanish population finding different behaviors in the evolution of dependency according to age and sex. These functional profiles or trajectories reflect the individual evolution across time of the dependency intensity (quantified by the dependency rating scale).

The standard way to estimate life expectancy is to use Cox regression model to obtain survival probabilities (see [4]). However, in our context, the event of interest is not 'survival' itself, but 'staying free of dependency at a given age'. We propose to use Cox regression model to obtain those probabilities and the estimation of LEFD is then straightforward. However, it must be pointed out that EDAD 2008 only contains records of alive people at 2008, hence the effect of death is ignored. That is, the estimated staying free of dependency probability at a given age is in fact the staying free of dependency probability at a given age given that a person is alive at that age. Then marginal probabilities are obtained by multiplying these estimates by survival probabilities given by mortality tables for the general population (in absence of specific mortality tables for dependent population).

We consider quantitative and categorical variables, such as the dependency rating scale value, sex and disabilities suffered. In fact, the dependency rating scale is of functional type, since it contains the trajectory of the evolution of dependency from 30 years to the end of the considered interval. However, in order to include this variable in the model with the other categorical variables, we will transform the trajectories into quantitative variables, for example, considering their distance to some given pattern. This pattern is obtained by means of functional tools, such as time-warping and functional depths, in order to take into account the time variability present in the dependency trajectories (see [2]).

## 3 Preliminary Results

To estimate LEFD we consider three different scenarios corresponding to the three degrees of dependency given by the Spanish legislation (I-moderate, II-severe, III-major). Indeed, for a non dependent person we calculate three different LEFDs, which are the expected number of years that a person can live out of each one of these dependency degrees.

In order to obtain a sample as homogeneous as possible, we include in the analysis people that in 2008 were between 70 and 80 years old, with a dependency rating scale value of 0 at the age of 30. As mentioned above, the variables included in the model are sex, the dependency trajectory and several disabilities suffered. In particular, the considered disabilities are the ability to perform postural changes, personal hygiene, relieve themselves, conduct normal activities of domestic life, maintain interpersonal relationships. Regarding the dependency trajectory, we quantify it by computing the $L^2$-distance from each curve to the deepest one with respect to the modified band depth (see [5]). In the time warping framework, that is, when a sample of curves exhibit high temporal variability (or horizontal shifting), which is the case for dependency trajectories, the modified band deepest curve has been proven to be a robust estimator of the underlying average pattern (see [3]). In order to calculate the deepest curve of a sample all the curves must be defined in the same time interval. Therefore, we partition the sample in groups of people of the same age (in years) and sex obtaining 22 subsamples. In each one of these subsamples we calculate the distance of each curve to the deepest one, yielding a numerical summary for each one of the trajectories. See Fig. 1 for an example of the deepest curve in one of the subsamples considered.

**Table 1** AIC[a] for two fitted models. Model A: sex, disabilities suffered and dependency path as explanatory variables; Model B: sex and disabilities suffered as explanatory variables

| Dependency degree | Model A | Model B | Relative difference |
|---|---|---|---|
| I-moderate | 6753202 | 7020290 | −4.0 % |
| II-severe | 4029275 | 4175027 | −3.6 % |
| III-major | 1656159 | 1670890 | −0.9 % |

[a] $AIC = 2k - 2\ln(L)$, where $k$ is the number of parameters, and $L$ is the maximized value of the likelihood

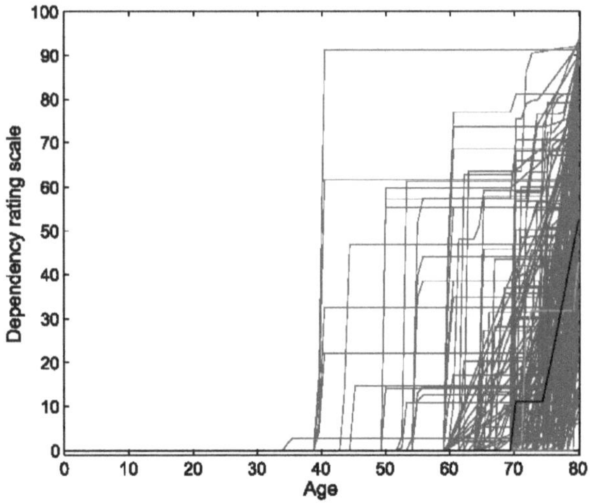


**Fig. 1** Dependency paths for women aged 80 and deepest curve of the sample in *black*

A possible way to assess whether the personal dependency curves help to estimate LEFD is to consider two models (with and without the quantitative information for the dependency paths) for each one of the three scenarios described above. In Table 1 we compare them with the Akaike Information Criterion, where we can see that the preferred model in all the scenarios is the one including the dependency trajectories. These preliminary results lead us to conclude that the historical personal information contained in the dependency path, and not only the current dependency status of a person, is relevant to estimate his/her future dependency situation.

## References

1. Albarrán, I., Alonso, P., Arribas-Gil, A.: Dependency evolution in Spanish disabled population: a functional data analysis approach. WP 13-04, Statistics and Econometric Series 03, Universidad Carlos III de Madrid (2013). http://hdl.handle.net/10016/16239
2. Arribas-Gil, A., Müller, H.-G.: Pairwise dynamic time warping for event data. Comput. Stat. Data Anal. **69**, 255–268 (2014)

3. Arribas-Gil, A., Romo, J.: Robust depth-based estimation in the time warping model. Biostatistics **13**, 398–414 (2012)
4. Cox, D.R., Oakes, D.: Analysis of Survival Data. Chapman and Hall, London (1984)
5. López-Pintado, S., Romo, J.: On the concept of depth for functional data. J. Am. Stat. Assoc. **104**, 486–503 (2009)

# Evaluation of Volatility Forecasts in a VaR Framework

**Alessandra Amendola and Vincenzo Candila**

**Abstract** Many methods can be considered to select which volatility model has a better forecast accuracy. In this work a loss function approach in a Value at Risk (VaR) framework is chosen. By using high-frequency data it is possible to achieve a consistent estimate of the VaR bootstrapping the intraday increments of an asset. The VaR estimate is used to find a threshold discriminating low from high loss function values. The analysis concerns the high-frequency data of a stock listed on the New York Stock Exchange.

**Keywords** Volatility · Value at Risk · Loss function · Bootstrap

## 1 Introduction

The evaluation of volatility forecasts produced by a set of competing models is generally carried out through a statistical (using a MSE, RMSE functions, for instance) or an economic approach. The economic approach evaluates the volatility predictions indirectly by using utility functions [3] or other risk measures like the Value at Risk [4]. The evaluation of volatility predictions through the Value at Risk (VaR) measures concerns some tests—like the Unconditional and Conditional Coverage (CC) tests [2]—about the occurrence, called violation, that the portfolio's loss is greater than the VaR. Unfortunately these tests suffer from low statistical power, as highlighted in [5].

The aim of this work is to investigate the opportunity to use the loss functions in a VaR framework in order to evaluate the volatility predictions of a set of competing models. Following [1], bootstrapping the intraday increments of a generic asset allows to have a consistent estimate of any characteristics of that asset's daily return. Hence the VaR measure is obtained as a quantile of the estimated distribution of

A. Amendola · V. Candila (✉)
Dept. of Economics and Statistics, University of Salerno, via Giovanni Paolo II 132, 84084 Fisciano (SA), Italy
e-mail: vcandila@unisa.it

A. Amendola
e-mail: alamendola@unisa.it

C. Perna, M. Sibillo (eds.), *Mathematical and Statistical Methods for Actuarial Sciences and Finance*, DOI 10.1007/978-3-319-05014-0_2,

the daily return. Then the performances of seven volatility models belonging to the family of the GARCH models are compared through two loss functions, of which one is a new proposed one.

The work is organized as follows. Section 2 presents the bootstrap method used to estimate the daily VaR measure. In Sect. 3 the two loss functions in a VaR framework are illustrated. Data description and the results of the volatility evaluation are in Sect. 4. Section 5 concludes.

## 2 Bootstrapping the Intraday Increments

Let $\chi_{t,N} = \{q_{t,1}, \ldots, q_{t,N}\}$ be a sequence of $N$ intraday increments for a generic day $t$ and a generic asset, such that $q_{t,n} = \log(P_{t,n}) - \log(P_{t,n-1})$, with $P_{t,n}$ denoting the observed intraday price. The open-to-close daily return $r_{t,N}$ is given by $r_{t,N} = \sum_{n=2}^{N} q_{t,n}$. Because of the dependence in the sequence $\chi_{t,N}$, the Stationary Bootstrap (SB) of [7] is used. In the SB, for each day $t$, $B$ re-sampled intraday sequences are calculated, each of length $N$. A re-sampled intraday sequence is formed by $N$ sampled blocks, whose average block length relies on the dependence exhibited within $\chi_{t,N}$. As [1], we use the procedure described in [6] in order to estimate the average block length. Once the bootstrapped sequence is obtained, the resulting summation represents a re-sampled daily return, independent of the original one, but generated by the same distribution, as $\{N, B\} \to \infty$. Hence any moment or quantile of the original return can be now estimated by means of this $B$ *i.i.d.* sequence. We focus on $\widehat{VaR}_t$, a consistent estimate of the 5 % VaR for the day $t$.

## 3 Loss Function in a VaR Framework

A loss function (LF) compares $r_{t,N}$ to the VaR measure. In this work we consider two loss functions, illustrated in Table 1. The first LF due to [5] is called Magnitude loss function (MLF). The second is the new proposed Asymmetric loss function (ALF), that penalizes more the models with an actual number of violations ($\hat{\alpha}$) greater than the expected one ($\alpha_0$). For the ALF, $P$ is the penalizing quantity that overpenalizes the model if $\hat{\alpha} > \alpha_0$. The average of the LS is called numerical score (*NS*) and it is denoted by $NS_M$ and $NS_A$ for the MLF and the ALF, respectively. Once the consistent estimate of the VaR has been obtained, it is possible to find a threshold that discriminates between low from high *NS*. The procedure is again based on the block bootstrap. The threshold is an empirical quantile of the distribution of the *NS*, when $\widehat{VaR}$ is used. If a *NS* of a volatility model lies above the threshold, that model is considered rejected.

**Table 1** Loss functions

| Magnitude loss function (MLF) | Asymmetric loss function (ALF) |
|---|---|
| $Loss_{M,t} = \begin{cases} 1 + (r_t - VaR_t)^2 & \text{if } r_t \leq VaR_t \\ 0 & \text{if } r_t > VaR_t \end{cases}$ | $Loss_{A,t} = \begin{cases} 1 + P \cdot (r_t - VaR_t)^2 & \text{if } r_t \leq VaR_t \\ 0 & \text{if } r_t > VaR_t \end{cases}$ |
| $NS_M = T^{-1} \sum_{t=1}^{T} Loss_{M,t}$ | $NS_A = T^{-1} \sum_{t=1}^{T} Loss_{A,t}$ |

**Table 2** Competing models

| | M1 | M2 | M3 | M4 | M5 | M6 | M7 |
|---|---|---|---|---|---|---|---|
| Method[a] | G(1, 1) | G(1, 1) | G(1, 1) | RM | GJR(1, 1) | GJR(1, 1) | GJR(1, 1) |
| $z_t \sim$[b] | $N(0, 1)$ | $t(v)$ | $sk - t(v, \xi)$ | – | $N(0, 1)$ | $t(v)$ | $sk - t(v, \xi)$ |

[a]G(1, 1) stands for GARCH(1, 1), RM for Riskmetrics

[b]$t(v)$ and $sk - t(v, \xi)$ represent $t$ and skewed-t distributions of the innovation vector $z_t$, such that $r_t = z_t h_t$

# 4 Empirical Analysis

The empirical analysis concerns the evaluation of the volatility predictions for a set of competing models, illustrated in Table 2.

The dataset[1] consists of the Capital One Financial Corporation one-minute trade prices. Once forecasted the conditional standard deviation, denoted by $\hat{h}_{t,m}$, with $m = 1, \ldots, 7$, we evaluate the performance looking at the MSE[2] and the exceedance of the threshold. The results of the comparison are showed in Table 3.[3]

If we only look at the MSE (statistical approach), the best model is M6, even though it exhibits more violations than expected, i.e. almost 6 % against the 5 %. Rows 3–4 show the performances of the models when the MLF is used. M7 is rejected because its *NS* lies above the threshold. Row 4 represents the ratio between the *NS* of each model and the *NS* when $\widehat{VaR}$ is used. The closer the ratio is to one, the better that model is. And the MLF awards M3. Unfortunately, there are too many models below the threshold. The situation becomes clearer if the ALF is used, as showed in rows 5–6. For the proposed LF, all the GJR models have *NS* above the threshold. The choice shrinks to models 1–4 and looking at the ratio (row 6), also the ALF awards M3. Finally, we argue that for the period considered and our mixed

[1]The data have been obtained from Tick Data, which is a provider of historical intraday market data. The sample period starts on April 8, 1997 and ends on December 31, 2003 (1695 trading days).

[2]The MSE is computed considering the distance between each $\hat{h}_{t,m}$ and the realized volatility, obtained summing the intraday returns at 5 minutes frequency and then it is multiplied by 1000.

[3]For brevity, we omit the results of the CC test because the null hypothesis—correct number of violations and independence—has been always rejected.

**Table 3** Forecasts comparison

| | M1 | M2 | M3 | M4 | M5 | M6 | M7 |
|---|---|---|---|---|---|---|---|
| MSE | 0.1523 | 0.1464 | 0.1484 | 0.1411 | 0.1346 | 0.1299 | 0.1351 |
| $\hat{\alpha}^{a}$ | 0.0457 | 0.0533 | 0.0526 | 0.0484 | 0.0567 | 0.0595 | 0.0630 |
| $NS_M < TR_M{}^{b}$ | Yes | Yes | Yes | Yes | Yes | Yes | No |
| $NS_M / NS_M{}^{c}$ | 0.8697 | 1.0145 | 1.0013 | 0.9224 | 1.0802 | 1.1329 | 1.1986 |
| $NS_A < TR_A$ | Yes | Yes | Yes | Yes | No | No | No |
| $NS_A / NS_A$ | 0.8724 | 1.0197 | 1.0061 | 0.9258 | 1.0865 | 1.1409 | 1.2083 |

[a] $\hat{\alpha}_m$ is the frequency of violations, for each model $m$

[b] $NS_M < TR$ is Yes if the numerical score of a model lies below the threshold

[c] $NS_M / NS$ is the ratio between the numerical score of a model and the numerical score obtained with $\widehat{VaR}$

approach M3, a GARCH(1, 1) model with $z_t \sim sk - t(\upsilon, \xi)$, is the best volatility model.

## 5 Conclusions

The evaluation of volatility forecasts by means of statistical or economic approaches may lead to ambiguous conclusions. We aimed to investigate the opportunity to use the loss function in a VaR framework. To do this, a consistent estimate of the VaR measures has been provided by using the stationary bootstrap. Then these VaR measures have been used to find the threshold discriminating low from high loss function values. The analysis has been conducted with two loss functions, of which one is new. It has emerged that this method helps the model selection in situations in which the traditional approaches do not clearly determine the best model.

## References


1. Bowers, C., Heaton, C.: Bootstrapping daily returns (2013). Available at SSRN http://dx.doi.org/10.2139/ssrn.2361393
2. Christoffersen, P.F.: Evaluating interval forecasts. Int. Econ. Rev. **39**(4), 841–862 (1998)
3. Fleming, J., Kirby, C., Ostdiek, B.: The economic value of volatility timing using "realized" volatility. J. Financ. Econ. **67**(3), 473–509 (2003)
4. Jorion, P.: Value at Risk. McGraw-Hill, New York (2007)
5. Lopez, J.A.: Methods for evaluating value-at-risk estimates. Econ. Policy Rev. **4**(3), 119–124 (1998)
6. Patton, A., Politis, D.N., White, H.: Correction to automatic block-length selection for the dependent bootstrap. Econom. Rev. **28**(4), 372–375 (2009)
7. Politis, D.N., Romano, J.P.: The stationary bootstrap. J. Am. Stat. Assoc. **89**, 1303–1313 (1994)

# Optimal Cut-Off Points for Multiple Causes of Business Failure Models

**Alessandra Amendola and Marialuisa Restaino**

**Abstract** In studies involving bankruptcy prediction models, since the attention is focused on the classification of firms into groups according to their financial status and the prediction of the status for new firms, optimal cutoff points have to be chosen. Some methods have been developed for two-group classification. Until now, there are few references on how to determine optimal thresholds when the groups are more than two. Here, a method based on the optimization of both correct classification rate and expected cost misclassification (ECM) is proposed for determining optimal cutoff points when there are multiple causes of business failure. The proposed procedure has been tested on a real data set.

**Keywords** Optimal cut-off points · ECM · $k \times k$ confusion table · Business failure

## 1 Introduction

Over the last decades the research studies on financial distress are essentially concerned with three aspects: the choice of the best model for identifying the failure process; the selection of the best set of covariates for classifying firms according to their status and their contribution to the performance of the model; and the evaluation of models' prediction power and ability. In this paper our attention has been focused on the third aspect.

The bankruptcy prediction typically involves the classification of firms in a group according to their financial status (bankruptcy, liquidation, merge, and so on). Therefore the accuracy of the classification can be evaluated by means of different measures, such as the correct classification rate, Type I error (or FP rate) and Type II error (or FN rate) [5, 6]. In order to compute these measures, as a prediction model

A. Amendola · M. Restaino (✉)
Department of Economics and Statistics, University of Salerno, via Giovanni Paolo II 132, 84084 Fisciano (SA), Italy
e-mail: mlrestaino@unisa.it

A. Amendola
e-mail: alamendola@unisa.it

C. Perna, M. Sibillo (eds.), *Mathematical and Statistical Methods for Actuarial Sciences and Finance*, DOI 10.1007/978-3-319-05014-0_3,

**Table 1** Confusion Matrix for three groups

| Actual Class | | *Class 0* | *Class 1* | *Class 2* | |
|---|---|---|---|---|---|
| Predicted class | *Class 0* | TP(0) | $E(0\|1)$ | $E(0\|2)$ | $C_0$[a] |
| | *Class 1* | $E(1\|0)$ | TP(1) | $E(1\|2)$ | $C_1$[a] |
| | *Class 2* | $E(2\|0)$ | $E(2\|1)$ | TP(2) | $C_2$[a] |
| | | $T_0$[b] | $T_1$[b] | $T_2$[b] | $T_0 + T_1 + T_2$ |

[a] $(C_0, C_1, C_2)$ are the units that are predicted to be in class 0, 1, 2, respectively.

[b] $(T_0, T_1, T_2)$ are the units for which the actual class is 0, 1, 2, respectively.

attributes a score to each firm in the data set, it is essential to determine optimal cut-off points, used for discriminating firms according to their status and for classifying firms into groups/classes.

For the binary data, some methods have been used, such as the minimization of the total number of misclassifications, the intersection of the distributions of the two groups, the minimization of the Type I error, and so on. The question is still open when the classes are more than two, because in this case more than one optimal cut-off point is needed.

The aim of this paper is to select optimal cut-off points when there are more than two groups and compute the accuracy measures, for evaluating the classification and prediction ability of a model.

## 2 Accuracy Measures and Optimal Cut-Offs

For evaluating the performance of an algorithm and/or a classifier and the quality of classification, different measures can be computed from a *confusion matrix* (also known as *classification table*) which records correctly and incorrectly cases for each class. The binary classification matrix can be extended to a multiple case when the groups are more than two. Since it becomes larger and larger at increasing the number of classes, the case of three groups has been considered for sake of simplicity.

Based on the Table 1, some accuracy measures can be computed, such as:

1. *Accuracy*: $Acc = \frac{TP(0)+TP(1)+TP(2)}{T_0+T_1+T_2}$;
2. *FP rate* (Type I error): $FPrate = \frac{E(0|1)+E(0|2)}{T_1+T_2}$;
3. *FN rate* (Type II error): $FNrate = \frac{E(1|0)+E(2|0)}{T_0}$.

In order to estimate these measures, in presence of two groups, one cut-off point is needed, while for the three classes two cut-off points have to be chosen: one for distinguishing between class 0 and class 1 and one for distinguishing between class 1 and class 2. Starting from the equation available for the binary data, the

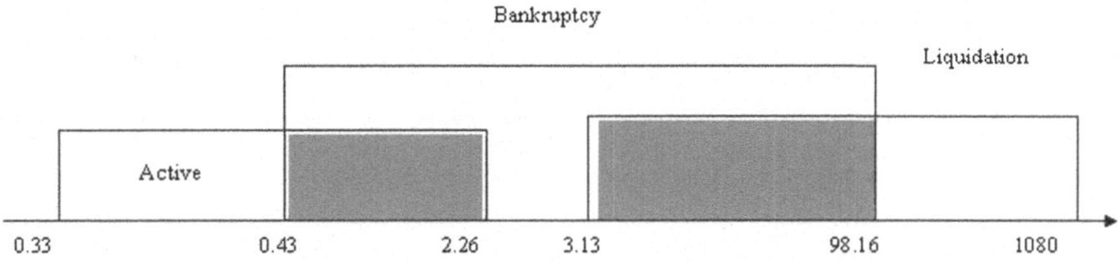


**Fig. 1** Range of risk scores for each event and the *grey zones*

minimization of the *Expected Cost of Misclassifications* (ECM) for the three groups is given by:

$$ECM = P_0 C_{1,2|0} \frac{n_0}{N_0} + P_1 C_{0,2|1} \frac{n_1}{N_1} + P_2 C_{0,1|2} \frac{n_2}{N_2} \quad (1)$$

where $(P_0, P_1, P_2)$ are the prior probabilities of group 0, 1 and 2, respectively; $(n_0, n_1, n_2)$ are the number of units misclassified for the group 0, 1 and 2, respectively; $(N_0, N_1, N_2)$ are the sample sizes for the three groups; $C_{1,2|0}$ is the cost of misclassifying an observation belonging to group 0 into groups 1 or 2; $C_{0,2|1}$ is the cost of misclassifying an observation belonging to group 1 into groups 0 or 2; $C_{0,1|2}$ is the cost of misclassifying an observation belonging to group 2 into groups 0 or 1.

## 3 Empirical Results

The financial data set considered here, drawn from Amadeus database of Bureau van Dijk, consists of building Italian firms that left the market between 2004 and 2010 for two main causes: bankruptcy and liquidation. The reference group is given by active firms. Starting from the financial statements of each firm included in the sample for a total of 38,028 balance sheets, we compute $nv = 16$ potential predictors, chosen among the most relevant in financial distress literature [2, 4]. Finally, the sample is divided into two parts: *in-sample set*, for evaluating the classification ability of a model, and *out-of-sample*, in order to estimate the prediction capability.

The model considered in the paper is the *competing risks model*, an extension of the mortality model for survival data. It is based on one transient state (alive state) and a certain number of absorbing states, corresponding to market exit from different causes. Thus, all transitions are from the state alive (for details, see [3]).

Since our aim is to determine threshold points which discriminate between the three groups (active, bankruptcy, liquidation), first of all we looked at the minimum and maximum of the risk score for each event and we noted that there are two grey zones in sense of [1] (Fig. 1). In order to classify the firms into three groups, we have to choose two cut-off points, according to two criteria: the accuracy is maximized and the FN rate is minimized; the ECM is minimized.

For this purpose, we divided the two grey zones into several subintervals, and for each one, the three measures (accuracy, FN rate and ECM) are computed. Looking

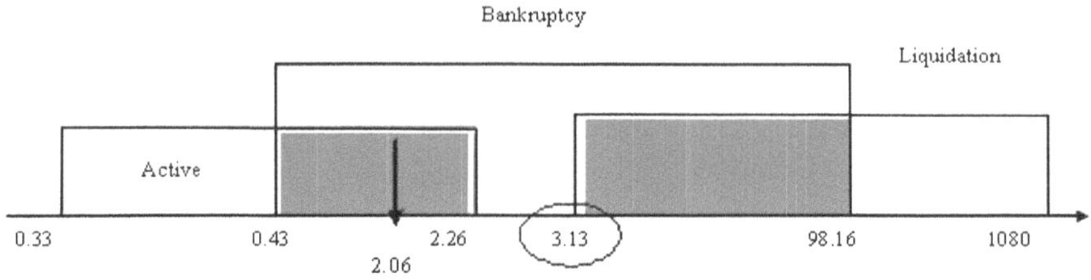


**Fig. 2** The chosen cut-off points of risk scores

**Table 2** Accuracy measures for the chosen cut-off points

| Cut-off Grey Zone1 | Cut-off Grey Zone2 | Accuracy | FN rate | ECM | Accuracy | FN rate | ECM |
|---|---|---|---|---|---|---|---|
| | | In-sample | | | Out-of-sample | | |
| 2.06 | 3.13 | 0.98782 | 0.00000 | 0.00052 | 0.99893 | 0.00000 | 0.00000 |

at the results,[1] we noted that the accuracy is increased at approaching the maximum of the both zones. Then, fixed the value of the second zone, the FN rate is always zero except when the maximum of the first grey area is reached. This result is stable independently of the values of the other zone. Finally, again fixed the values of the second range, and looking at the values of ECM, we observed that it is decreased at approaching the maximum of the first area. Now, by varying the values of the second zone, the ECM is increased when the maximum is reached.

Therefore the cut-off points that respect all the criteria are: 2.05 for the first grey zone and 3.13 for the second one (Fig. 2).

After choosing the two thresholds, we computed the accuracy measures for in-sample and out-of-sample (Table 2) and it can be observed that the performance of the model is good in terms of both classification ability and prediction capability.

In summary, in this paper we proposed some criteria for identifying optimal cut-off points in presence of three groups and we compared the performance of these methods. This approach can be easily extended to the case of $k$ groups.

## References

1. Altam, E.I.: Financial ratios, discriminant analysis and the prediction of corporate bankruptcy. J. Finance **23**, 589–609 (1968)
2. Altman, E.I.: Predicting financial distress of companies: revisiting the Z-score and ZTM model. Working Paper, New York University (2000)
3. Andersen, P.K., Abildstrøm, S.Z., Rosthøj, S.: Competing risks as a multi-state model. Stat. Methods Med. Res. **11**, 203–215 (2002)
4. Dimitras, A., Zanakis, S., Zopudinis, C.: A survey of businesses failures with an emphasis on failure prediction methods and industrial applications. Eur. J. Oper. Res. **27**, 337–357 (1996)

[1] The tables of results are available upon requests from the authors.

5. Koh, H.C.: The sensitivity of optimal cutoff points to misclassification costs of type I and type II errors in going-concern prediction context. J. Bus. Finance Account. **19**(2), 187–197 (1992)
6. Nanda, S., Pendharkar, P.: Linear models for minimizing misclassification costs in bankruptcy prediction. Int. J. Intell. Syst. Account. Finance Manag. **10**, 155–168 (2001)

# Maximum Empirical Likelihood Inference for Outliers in Autoregressive Time Series

**Roberto Baragona, Francesco Battaglia, and Domenico Cucina**

**Abstract** Outliers in time series are usually distinguished in additive, innovation, and transient and permanent change. An approach based on empirical likelihood is presented for estimating outliers of the four types in a linear autoregressive time series. Theoretical results are illustrated along with hints for future research.

**Keywords** Additive and innovation outlier · Level change · Confidence intervals

## 1 Introduction

Classification of outliers in time series in the four groups of additive (AO), innovation (IO), transient (TC) and permanent (LC) level change has been introduced by [9].

Let $y = (y_1, y_2, \dots, y_n)'$ be the observed time series and

$$y_t = c_t + z_t \tag{1}$$

the basic outlier model, where $c_t$ is a deterministic function that represents outliers, and $z_t$ is the unobserved outlier free time series. Let $z_t$ follow an autoregressive process $\Phi(B)z_t = \varepsilon_t$ where $\Phi(B) = 1 - \phi_1 B - \phi_2 B^2 \cdots - \phi_p B^p$ is the autoregressive polynomial and $\{\varepsilon_t\}$ is a sequence of independent identically distributed random variables with mean zero and finite variance. An AO at time $q$ is defined by letting in (1) $c_t = \omega I[t = q]$ where $I[.]$ is the indicator function. An IO at time $q$ is obtained by letting $c_t = \Phi(B)^{-1}\omega I[t = q]$. Level changes are defined, the LC

R. Baragona
Lumsa University of Rome, Rome, Italy
e-mail: r.baragona@lumsa.it

F. Battaglia (✉) · D. Cucina
Sapienza University of Rome, Rome, Italy
e-mail: francesco.battaglia@uniroma1.it

D. Cucina
e-mail: domenico.cucina@uniroma1.it

C. Perna, M. Sibillo (eds.), *Mathematical and Statistical Methods for Actuarial Sciences and Finance*, DOI 10.1007/978-3-319-05014-0_4,

by assuming in (1) $c_t = \frac{1}{1-B}\omega I[t=q]$, the TC by letting $c_t = \frac{1}{1-\delta B}\omega I[t=q]$ for some $\delta \in (0, 1)$.

The usual steps of outlier treatment are a detection step, where candidate outliers are located, a test step where the null hypothesis of no outliers is checked against the alternative that the located observation is an outlying one, and a step for estimating the outlier size, in view of using such estimate for time series adjustment and correction. Common approaches refer for instance to the likelihood ratio test ([9]), methods based on influence ([7]), genetic algorithms ([1]).

Empirical likelihood (EL) methods ([6]) allow maximum likelihood hypothesis testing in the framework of a distribution free model specification for outliers. Originally derived for building confidence intervals for the mean of a random sample without assumptions on the form of the probability distribution, the EL methods have been extended to linear models and time series (e.g. [3, 4]), and recently to breaks in regression models ([2]).

The maximum EL ratio test is generally used in applications when inference on a firm theoretical ground is needed but there is not enough information for assuming a family of probability distributions that may fit well the data. In the same view the maximum EL framework is suggested here for outlier size inference without assumptions on some underlying probability distribution.

The rest of the paper is organized as follows. Section 2 introduces the EL methods, Sect. 3 shows their applications for inference about outliers in time series, in Sect. 4 conclusions are drawn.

## 2 Empirical Likelihood Methods

Let $(x_1, x_2, \ldots, x_n)$ denote an observed random sample from the random variable $x$ with $E(x) = \mu$ and finite variance. That sample may be thought of as drawn from a discrete probability distribution concentrated only at the values $x_1, x_2, \ldots, x_n$ with probabilities $p_i$ ($p_i > 0, \sum p_i = 1$), called the empirical distribution.

In this case the probability of getting exactly that sample would be $p_1 p_2 \ldots p_n$. The mean of that distribution is $\mu = \sum_i p_i x_i$, therefore the possible discrete distributions generating the observed sample, given $\mu$, are the set $\{(p_1, p_2, \ldots, p_n) : p_i > 0, \sum p_i = 1, \sum p_i(x_i - \mu) = 0\}$ and the largest probability to get the observed sample is

$$EL(\mu) = \max_{p_1,\ldots,p_n} \left\{ \prod_{i=1}^{n} p_i : p_i > 0, \sum_{i=1}^{n} p_i = 1, \sum_{i=1}^{n} p_i(x_i - \mu) = 0 \right\}. \quad (2)$$

This is called the EL profile for $\mu$. The maximum is reached for $\mu = \bar{x} = \sum x_i/n$ and equals $n^{-n}$, corresponding to $p_i = 1/n$, $i = 1, \ldots, n$. Thus, $\bar{x}$ is the maximum EL estimator (MELE) of $\mu$.

This approach may be generalized by using the estimating equations (e.g. [6, 8]) to replace the constraint $E(x - \mu) = 0$ in Eq. (2). Let $\theta = (\theta_1, \ldots, \theta_p)'$ denote the parameter vector to be estimated using the observed data $x = (x_1, x_2, \ldots, x_n)'$. The relationship between the parameters $\theta$ and the data may be summarized by $p$ estimating functions $g_k(x, \theta)$ chosen so that the $p$ estimating equations $E\{g_k(x, \theta)\} = 0$, $k = 1, \ldots, p$ determine $\theta$ uniquely. The maximized EL ratio (ELR) profile is obtained by dividing (2) by its unconstrained maximum $n^{-n}$ and replacing $x_i - \mu$ with $\{g_1(x_i, \theta), \ldots, g_p(x_i, \theta)\}$.

Let $\ell(\theta) = -2\log\{\text{ELR}(\theta)\}$ be the maximized empirical log likelihood ratio profile and assume that $\theta$ is the true parameter value, i.e. the unique solution of $E\{g_k(x, \theta)\} = 0$, $k = 1, 2, \ldots, p$. Then it may be shown ([5, 8]) that under some regularity conditions, as $n \to \infty$, $\ell(\theta)$ converges in distribution to a $\chi^2_p$.

## 3 Empirical Likelihood for Outlier Estimation

Let the outlier be located at time $q$ and be $\omega$ its size. Then Eq. (1) may be rewritten as

$$y_t = \sum_{j=1}^{p} (y_{t-j} - c_{t-j}\omega)\phi_j + c_t\omega + \varepsilon_t. \tag{3}$$

The parameter vector is $\theta = (\phi_1, \ldots, \phi_p, \omega)'$ and its length is $p + 1$. The consideration of $\omega$ as an additional parameter makes model (3) a non linear time series model of the form $y_t = f(x_t, \theta) + \varepsilon_t$, where $x_t = (y_{t-1}, y_{t-2}, \ldots, y_{t-p})$.

A MELE may be obtained along the guidelines outlined in Sect. 2. The estimating equations may be taken equal the $p + 1$ derivatives with respect to the parameters of the residual sum of squares $\sum_t [y_t - f(x_t, \theta)]^2$:

$$g_k(x_t, \theta) = (c_{t-k}\omega - y_{t-k})\varepsilon_t(\theta), \quad k = 1, \ldots, p,$$

$$g_{p+1}(x_t, \theta) = \left(\sum_{j=1}^{p} c_{t-j}\phi_j - c_t\right)\varepsilon_t(\theta)$$

where $\varepsilon_t(\theta) = y_t - c_t\omega - \sum_{j=1}^{p}(y_{t-j} - c_{t-j}\omega)\phi_j$. Let $\hat{\theta} = (\hat{\phi}_1, \ldots, \hat{\phi}_p, \hat{\omega})'$. The MELE estimate is defined as

$$\hat{\theta} = \arg\max_{\theta}\left\{\max_{p_1,\ldots,p_n}\left(\prod_{i=1}^{n} p_i : p_i > 0, \sum_{i=1}^{n} p_i = 1, \sum_{i=1}^{n} p_i\big(g_k(x_i, \theta)\big) = 0\right)\right\}$$

and therefore coincides with the least squares estimate. Since $\ell(\theta) \to \chi^2_{p+1}$ in distribution when $\theta$ is the true value, confidence regions for the parameter $\theta$ may be obtained as:

$$\{\theta : \ell(\theta) < \overline{\chi}^2_{p+1,1-\alpha}\}$$

where $\overline{\chi}^2_{p+1,1-\alpha}$ is the $(1 - \alpha)$ quantile of the chi square distribution with $p + 1$ degrees of freedom. Confidence intervals for the outlier size may also be obtained based on Corollary 5 of [8]. Let $\omega$ be the true value of the outlier size, and denote by $\tilde{\phi}(\omega)$ the MELE of the autoregressive parameters given $\omega$ (simply obtained from $\sum_t g_k(x_t, \phi_1, \ldots, \phi_p, \omega) = 0$, $k = 1, \ldots, p$ for $\omega$ fixed); then $\ell(\tilde{\phi}(\omega), \omega)$ converges to a $\chi_1^2$ distribution. It suggests the following confidence interval for $\omega$:

$$\left\{\omega : \ell\left(\tilde{\phi}(\omega), \omega\right) < \overline{\chi}^2_{1,1-\alpha}\right\}.$$

## 4 Conclusions

Empirical likelihood methods provide a convenient framework for developing confidence regions for outliers in time series. Likewise, tests for the presence of outliers in time series may be built. Specification of a probability distribution for the data is not needed but hypothesis testing develops along the same guidelines as for parametric likelihood. However, care is needed because not all values of the constants $\{c_t\}$ ensure conditions for convergence of $\ell(\theta)$ to a chi square distribution. Future research may be concerned with extension to other outlier types, patches and variance change for instance, and to a wider class of models such as general autoregressive moving average models and non linear autoregressive models.

**Acknowledgements** This research was supported by the grant C26A1145RM of the Universitá di Roma La Sapienza, and the national research PRIN2011 "Forecasting economic and financial time series: understanding the complexity and modelling structural change", funded by Ministero dell'Istruzione dell'Universitá e della Ricerca.

## References

1. Baragona, R., Battaglia, F., Calzini, C.: Genetic algorithms for the identification of additive and innovation outliers in time series. Comput. Stat. Data Anal. **37**, 1–12 (2001)
2. Baragona, R., Battaglia, F., Cucina, D.: Empirical likelihood for break detection in time series. Electron. J. Stat. **7**, 3089–3123 (2013)
3. Chan, N.H., Peng, L., Zhang, D.: Empirical-likelihood based confidence intervals for conditional variance in heteroskedastic regression models. Econom. Theory **27**, 154–177 (2011)
4. Chuang, C.-S., Chan, N.H.: Empirical likelihood for autoregressive models, with applications to unstable time series. Stat. Sin. **12**, 387–407 (2002)
5. Mykland, P.A.: Dual likelihood. Ann. Stat. **23**, 396–421 (1995)
6. Owen, A.B.: Empirical Likelihood. Chapman & Hall/CRC, Boca Raton (2001)
7. Peña, D.: Influential observations in time series. J. Bus. Econ. Stat. **8**, 235–241 (1990)
8. Qin, J., Lawless, J.: Empirical likelihood and general estimating equations. Ann. Stat. **22**, 300–325 (1994)
9. Tsay, R.S.: Outliers, level shifts and variance changes in time series. J. Forecast. **7**, 1–20 (1988)

# The Role of Fund Size and Returns to Scale in the Performance of Mutual Funds

**Antonella Basso and Stefania Funari**

**Abstract** In this contribution we investigate the effects of the size of mutual funds on their performance by using a Data Envelopment Analysis (DEA) approach. We discuss the role of fund size in the performance evaluation and wonder whether it is appropriate to include size information among the input/output variables of the DEA models. Moreover, we analyze the nature of returns to scale in mutual fund performance and investigate whether returns to scale are constant, increasing or decreasing in a set of European mutual funds.

**Keywords** Mutual fund performance evaluation · DEA · Size · Returns to scale

## 1 Introduction

This contribution addresses two issues that can be of interest when we evaluate the performance of mutual funds using a Data Envelopment Analysis (DEA) methodology, a nonparametric approach to efficiency analysis that in the last years have been increasingly used also to evaluate the performance of mutual funds.

First of all, we tackle the problem of investigating the effects of the size of mutual funds on their performance. This issue is particularly relevant in the case of variable returns to scale (VRS) models, but it has been considered only in a small number of papers in the DEA literature (see for instance [3, 4]). First, we discuss the role of fund size in the performance evaluation and wonder whether it is appropriate to include size information among the input/output variables of the DEA models. The question of the utility of including size in the performance evaluation arises especially when the analysis is focused on the point of view of financial investors.

A. Basso
Department of Economics, Ca' Foscari University of Venice, Cannaregio 873, 30121 Venice, Italy
e-mail: basso@unive.it

S. Funari (✉)
Department of Management, Ca' Foscari University of Venice, Cannaregio 873, 30121 Venice, Italy
e-mail: funari@unive.it

C. Perna, M. Sibillo (eds.), *Mathematical and Statistical Methods for Actuarial Sciences and Finance*, DOI 10.1007/978-3-319-05014-0_5,

Indeed, investors would like to maximize the financial results without being exposed to high risk levels, and may not care much about fund size.

A second issue concerns the nature of returns to scale in mutual fund performance. By considering a set of European mutual funds, we investigate whether returns to scale are constant, increasing or decreasing. The measure of fund efficiency is computed by using a suitable DEA model that has been recently proposed in the literature to assess fund performance ([2]); this model allows to take into consideration the main elements of an investment in mutual funds and is applicable also when the technology exhibits variable returns to scale (VRS).

In the empirical analysis we investigate the presence of economies or diseconomies of scale and investigate if size causes any significant effects on fund performance.

## 2 What About Fund Size?

From the point of view of a financial investor, two funds with identical features in terms of mean return, risk level, initial and exit fees can be considered as equivalent: if the values of these variables are identical, the two funds are perceived as equivalent by investors. This does not depend on the fund size in terms of total market value. Of course, if the aim were to evaluate the skill of fund managers, the changes in the size of the funds might well be relevant.

On the other hand, if we include size among the input variables, two funds with identical values of mean return, risk level and initial and exit fees but different size could have different performance scores. In this regard, let us consider the instance illustrated in Table 1, in which funds 11th to 20th have the same values of mean return, risk level, initial and exit fees and ethical level as funds 11–20, but they have three times the size.

Hence, from the investors' point of view, fund 1 is judged equivalent to fund 11, fund 2 is judged equivalent to fund 12, and so on.

Let us consider the fund size $S_j$ and the beta coefficient $\beta_j$ as inputs, and as output the following measure of the fund profitability: $M_{Sj} = S_j(1-c_{Ij})(1+R_j)^3(1-c_{Ej})$ (the final value after three years, net of initial ($c_{Ij}$) and exit ($c_{Ej}$) fees, computed on the fund size according to the annual rate of return $R_j$). The last column of table 1 shows the DEA score obtained with a BCC (named after Banker, Charnes, and Cooper [1]) model exhibiting VRS. As can be seen, the DEA score of fund $j$ is not always the same as that of fund $j+10$ (for example the score of fund 3 is not the same as that of fund 13). This shows that the performance scores computed in this way exhibit a bias. For this reason, even if in the DEA literature we can find a few contributions that take size into account (see for example [5] and [6]), we deem correct not to consider it and do not insert fund size among the inputs of the DEA models.

**Table 1** Data of the instance on the effect of size on the fund performance

| Fund | $\beta$ | $S$ | $c_I$ | $c_E$ | $R$ | $M_S$ | DEA score |
|---|---|---|---|---|---|---|---|
| Fund 1 | 1.60 | 5 | 0.050 | 0.00 | 0.06 | 5.657 | 1.000 |
| Fund 2 | 2.00 | 12 | 0.040 | 0.00 | 0.08 | 14.512 | 1.000 |
| Fund 3 | 1.20 | 34 | 0.030 | 0.01 | 0.04 | 36.727 | 0.978 |
| Fund 4 | 0.80 | 20 | 0.030 | 0.00 | −0.01 | 18.824 | 1.000 |
| Fund 5 | 0.80 | 21 | 0.025 | 0.00 | 0.02 | 21.728 | 1.000 |
| Fund 6 | 0.90 | 40 | 0.030 | 0.00 | −0.08 | 30.213 | 0.711 |
| Fund 7 | 0.92 | 22 | 0.000 | 0.03 | −0.03 | 19.476 | 0.835 |
| Fund 8 | 1.25 | 15 | 0.020 | 0.01 | −0.01 | 14.121 | 0.871 |
| Fund 9 | 1.10 | 5 | 0.025 | 0.00 | −0.10 | 3.554 | 1.000 |
| Fund 10 | 1.00 | 13 | 0.020 | 0.00 | −0.04 | 11.272 | 0.877 |
| Fund 11 | 1.60 | 15 | 0.050 | 0.00 | 0.06 | 16.972 | 0.974 |
| Fund 12 | 2.00 | 36 | 0.040 | 0.00 | 0.08 | 43.536 | 1.000 |
| Fund 13 | 1.20 | 102 | 0.030 | 0.01 | 0.04 | 110.181 | 1.000 |
| Fund 14 | 0.80 | 60 | 0.030 | 0.00 | −0.01 | 56.471 | 0.910 |
| Fund 15 | 0.80 | 63 | 0.025 | 0.00 | 0.02 | 65.185 | 1.000 |
| Fund 16 | 0.90 | 120 | 0.030 | 0.00 | −0.08 | 90.639 | 1.000 |
| Fund 17 | 0.92 | 66 | 0.000 | 0.03 | −0.03 | 58.429 | 0.839 |
| Fund 18 | 1.25 | 45 | 0.020 | 0.01 | −0.01 | 42.362 | 0.855 |
| Fund 19 | 1.10 | 15 | 0.025 | 0.00 | −0.10 | 10.662 | 0.679 |
| Fund 20 | 1.00 | 39 | 0.020 | 0.00 | −0.04 | 33.815 | 0.807 |

## 3 What About Returns to Scale?

In order to study returns to scale, we apply the DEA-V model proposed in [2], that considers the most significant variables for an investor with a well diversified portfolio and can be used even in the presence of negative mean returns. This is an output oriented BCC model with VRS and its dual version can be written as follows:

$$\max \quad z_0 + \varepsilon s_1^+ + \varepsilon s_1^- + \varepsilon s_2^- \tag{1}$$

$$\text{s.t.} \quad M_o z_0 - \sum_{j=1}^{n} M_j \lambda_j + s_1^+ = 0 \tag{2}$$

$$\sum_{j=1}^{n} K_j \lambda_j + s_1^- = K_o \tag{3}$$

$$\sum_{j=1}^{n} \beta_j \lambda_j + s_2^- = \beta_o \tag{4}$$

$$\sum_{j=1}^{n} \lambda_j = 1 \quad (5)$$

$$\lambda_j \geq 0 \quad j = 1, 2, \ldots, n \quad (6)$$

$$s_1^+, s_1^-, s_2^- \geq 0 \quad (7)$$

where $K_j = 1/(1 - c_{Ij})$ is the capital invested in fund $j$, net of the initial fees, $M_j = (1 + R_j)^3(1 - c_{Ej})$ is the final value of the investment net of the exit fee; $z_0$, $\lambda_j$ $(j = 1, 2, \ldots, n)$, $s_1^+$, $s_1^-$, $s_2^-$ are the dual variables associated to the constraints of the linear programming problem which is the primal of problem (1)–(7), $\varepsilon$ is a non-Archimedean constant and $o$ denotes the fund which is being evaluated.

We use the same database of European mutual funds presented in [2] (279 funds in the period June 2006–June 2009). Interestingly, the results indicate that half of the efficient funds exhibits constant returns to scale, while for the funds with VRS the returns to scale are increasing. Moreover, for the inefficient funds, if we consider the projection on the efficient frontier, the returns to scale are constant in 47 % of cases and increasing in 53 %.

## 4 Any Empirically Verifiable Effect of Size?

In order to see if there is an effect of fund size on performance, we have also tried to investigate if there is a significant influence of the fund size on the DEA performance score on the set of European funds analyzed in Sect. 3; to this aim we use a method similar to the one used by [3]. The fund size is measured by the *total market value* (expressed in millions of euro), and we compute the correlation coefficient between the fund size and the DEA performance scores, as well as its statistical significance. The performance scores are computed with model (1)–(7), while we consider fund size as an external variable. This choice seems natural, since we have seen in Sect. 2 that the direct inclusion of size among the variables of the model may lead to distorted results when we adopt the point of view of investors. Notice that also other contributions, such as [4], study the effects of scale on mutual funds performance by considering the funds size as external-environmental variable, not directly included in the DEA model.

The results of the investigation carried out indicate that the correlation coefficient between the fund size and the DEA performance scores is low and is not statistically different from 0 for all usual significance levels, so that there seems to be no significant correlation between size and performance. This conclusion is similar to that reached in [3] and supports the choice of omitting the fund size in the DEA models.

## References

1. Banker, R., Charnes, A., Cooper, W.W.: Some models for estimating technical and scale inefficiencies in data envelopment analysis. Manag. Sci. **30**, 1078–1092 (1984)

2. Basso, A., Funari, S.: Constant and variable returns to scale DEA models for socially responsible investment funds. Eur. J. Oper. Res. (2013). doi:10.1016/j.ejor.2013.11.024
3. Choi, Y.K., Murthi, B.P.S.: Relative performance evaluation of mutual funds: a non-parametric approach. J. Bus. Finance Account. **28**, 853–876 (2001)
4. Daraio, C., Simar, L.: A robust nonparametric approach to evaluate and explain the performance of mutual funds. Eur. J. Oper. Res. **175**, 516–542 (2006)
5. Haslem, J.A., Scheraga, C.A.: Data envelopment analysis of Morningstar's small-cap mutual funds. J. Invest. **15**, 87–92 (2006)
6. Pendaraki, K.: Mutual fund performance evaluation using data envelopment analysis with higher moments. J. Appl. Finance Bank. **2**, 97–112 (2012)

# A Robustness Analysis of Least-Squares Monte Carlo for R&D Real Options Valuation

**Marta Biancardi and Giovanni Villani**

**Abstract** In this paper we study the robustness of Least Squares Monte Carlo (LSM) in valuing R&D investment opportunities. As it is well known, R&D projects are characterized by sequential investments and therefore they can be considered as compound option involving a set of interacting American-type options. The basic Monte Carlo simulation takes a long time and it is computationally intensive and inefficient.

In this context, LSM method is a powerful and flexible tool for capital budgeting decisions and for valuing R&D investments. In particular way, stress testing different basis functions, we show the major technical advantages as reduction of the execution time and improvement in the simulation on the R&D projects valuation.

**Keywords** Least-squares Monte Carlo · R&D real options · Robustness analysis

## 1 Introduction

Option pricing theory has been used successfully in practice to value firm's investment opportunities. In real options, the options involve real assets as opposed to financial ones. To have a real option means to have the possibility for a certain period to either choose to realize an investment or to delay it, waiting better information. In particular, the R&D projects have received great attention in recent years, because these projects are similar to the purchase of an option on a future investments. An R&D investment usually involves several phases, in which the start of a phase depends on the success of the preceding phase. Therefore, the R&D investments can be considered as compound options. Moreover, in order to value the managerial flexibility to realize an investment before the maturity date, most of R&D investments are structured as compound American options. In this context, Monte Carlo

M. Biancardi · G. Villani (✉)
Department of Economics, Largo Papa Giovanni Paolo II 1, Foggia, Italy
e-mail: giovanni.villani@unifg.it

M. Biancardi
e-mail: marta.biancardi@unifg.it

C. Perna, M. Sibillo (eds.), *Mathematical and Statistical Methods for Actuarial Sciences and Finance*, DOI 10.1007/978-3-319-05014-0_6,

simulation, and in particular way the Least-Squares Monte Carlo (LSM) approach proposed by [3], is an attractive tool to solve complex real options models, as it is witnessed in [2] and [5].

The focus of our paper is to study the robustness of LSM in valuing R&D investment opportunities. In particular way, stress testing different basis functions, we show the major technical advantages as reduction of the execution time and improvement in the simulation on the R&D projects valuation.

## 2 The Basic Model

In this model, we assume a two-stage R&D investment which structure is the following:

- $R$ is the Research investment spent at initial time $t_0 = 0$;
- $IT$ is the Investment Technology to research innovations payed at time $t_1$. We further suppose that $IT = qD$ is a proportion $q$ of asset $D$, so it follows the same stochastic process of $D$;
- $D$ is the Development investment that the firm needs to invest to receive the R&D project's value. We assume that $D$ can be realized between $t_1$ and $T$;
- $V$ is the R&D project value.

Equations (1), (2) and (3) describe the evolutions of assets $V$ and $D$:

$$\frac{dV}{V} = (\mu_v - \delta_v)dt + \sigma_v dZ_t^v \tag{1}$$

$$\frac{dD}{D} = (\mu_d - \delta_d)dt + \sigma_d dZ_t^d \tag{2}$$

$$cov\left(\frac{dV}{V}, \frac{dD}{D}\right) = \rho_{vd}\sigma_v\sigma_d\, dt. \tag{3}$$

In particular way, investing $R$ at time $t_0$, the firm obtains a first investment opportunity that we can be value as a Compound American Exchange Option (CAEO) denoted by $C(S_k, IT, t_1)$. This option allows to realize the Investment Technology $IT$ at time $t_1$ and to obtain, as underlying asset, the option to realize the market launch; let denote by $S_k(V, D, T - t_1)$ this option value at time $t_1$ with maturity date $T - t_1$ and exercisable $k$ times. In detail, during the market launch, the firm has got a second investment opportunity to invest $D$ between $t_1$ and $T$ and to receive the R&D project value $V$. In particular way, using the LSM model, the firm must decide at any discrete time $\tau_k = t_1 + k\Delta t$, for $k = 1, 2, \ldots, h$ with $\tau_h = t_1 + h\Delta t = T$, whether to invest $D$ or to wait, and so to delay the decision at next time. In this way we capture the manager flexibility to invest $D$ before the maturity $T$ and so to realize the R&D cash flows.

After some manipulation, Eq. (4) describes the evolution of asset $P = \frac{V}{D}$ under the risk-neutral probability measure $\widetilde{Q}$:

$$P(t) = P_0 \exp\left\{\left(\delta_d - \delta_v - \frac{\sigma^2}{2}\right)t + \sigma Z^p(t)\right\} \tag{4}$$

**Table 1** Input values for R&D valuation

| Project | | I | II | III | IV |
|---|---|---|---|---|---|
| *R&D* Project Value | $V_0$ | 250 000 | 210 000 | 750 000 | 410 000 |
| Development Cost | $D_0$ | 140 000 | 200 000 | 950 000 | 310 000 |
| Investment Technology | $IT_0$ | 70 000 | 120 000 | 171 000 | 46 500 |
| Research Investment | $R$ | 50 000 | 40 000 | 35 000 | 100 000 |
| Exchange Comp. ratio | $q$ | 0.50 | 0.60 | 0.18 | 0.15 |
| Dividend-Yield of $V$ | $\delta_v$ | 0.20 | 0.15 | 0.15 | 0.15 |
| Dividend-Yield of $D$ | $\delta_d$ | 0.05 | 0.05 | 0 | 0 |
| Time to Maturity | $t_1$ | 1 year | 1 year | 2 year | 3 year |
| Time to Maturity | $T$ | 2 year | 3 year | 5 year | 7 year |
| Correlation | $\rho_{vd}$ | 0.38 | 0.26 | 0.08 | 0.12 |
| Volatility of $V$ | $\sigma_v$ | 0.83 | 0.64 | 0.54 | 0.88 |
| Volatility of $D$ | $\sigma_d$ | 0.32 | 0.41 | 0.15 | 0.31 |

where $\sigma = \sqrt{\sigma_v^2 + \sigma_d^2 - 2\sigma_v\sigma_d\rho_{vd}}$ and $Z^p$ is a geometric Brownian motion under $\widetilde{Q}$. The value of CAEO can be determined as the expected value of discounted cash-flows:

$$C(S_k, IT, t_1) = D_0 e^{-\delta_d t_1} E_{\widetilde{Q}}\big[\max\big(S_k(P_{t_1}, 1, T - t_1) - q, 0\big)\big].$$

The main contribution of LSM method is to determine the expected continuation values by regressing the subsequent discounted cash flows on a set of basis functions of current state variables. Unlike our previous paper (see [5]) , we show the effect on CAEO prices of a change in the type or the number of basis functions. As described in [1] and [4], the common choice of basis functions are the weighted Power, Laguerre, Hermite, Legendre, Chebyshev, Gegenbauer and Jacobi polynomials.

## 3 Expected Numerical Results on the Robustness of LSM

For complex options, such as the CAEO, the choice of basis functions is not clear, since we have to combine polynomials with other functions. The choice of polynomial family and the number of polynomials can produce results incorrect about the estimation of the continuation value of the option and so very inaccurate option value estimation. Table 1 summarizes the input parameters about four R&D investments that we are going to consider. We can observe that, as R&D present a high uncertainty about their results, we assume that $\sigma_v$ changes between 0.54 and 0.88.

For the CAEO option, following an early analysis, we find that the type and number of basis functions can affect both option prices and the computation time and, therefore, the choice among several R&D projects. An early test shows that, for the CAEO pricing, the accuracy improvement stops at 5 polynomials. The Standard Error $\varepsilon_n = \frac{\hat{\sigma}}{n}$ is a measure of simulation accuracy and it is estimated as the realised standard deviation of simulations divided by the square root of simulations. We assume that the number of paths is $m = 20\,000$, $n = 10\,000$ with $x = 20$ steps for year.

Moreover, the accuracy of simulation, and so the choice of basis function, is strong affected when the R&D projects are characterized by high volatility (such as Projects I and IV) and high time maturity (such Projects III and IV). In order to decrease the standard error it is necessary to increasing the number of paths, and so the computation time, or by using variance reduction procedures, such as the antithetic variates method.

Another result underlines that, when time is relevant, the Powers polynomial is one the fastest to compute. To give an idea of CFU computation time, it takes about three hours for basic MC and two hours for LSM using the three weighted Powers polynomial. So, from an early test, the results on robustness show that it is possible to increase the accuracy of LSM option valuation method without a major computational costs.

## References

1. Areal, N., Rodrigues, A., Armada, M.J.R.: Improvements to the least squares Monte Carlo option valuation method. Rev. Deriv. Res. **11**(1–2), 119–151 (2008)
2. Cortelezzi, F., Villani, G.: Valuation of R&D sequential exchange options using Monte Carlo approach. Comput. Econ. **33**, 209–236 (2009)
3. Longstaff, F.A., Schwartz, E.S.: Valuing American options by simulation: a simple least-squares approach. Rev. Financ. Stud. **14**(1), 113–147 (2001)
4. Moreno, M., Navas, J.F.: On the robustness of Least-squares Monte Carlo (LSM) for pricing American derivatives. Rev. Deriv. Res. **6**(2), 107–128 (2003)
5. Villani, G.: Valuation of R&D investment opportunities using the least-squares Monte Carlo method. In: Corazza, M., Pizzi, C. (eds.) Mathematical and Statistical Methods for Actuarial Sciences and Finance, pp. 287–299. Springer, Basel (2014)

# The Common Pool Problem of Intergovernmental Interactions and Fiscal Discipline: A Stackelberg Approach

**Giovanna Bimonte and Pietro Spennati**

**Abstract** In common pool models fiscal outcomes are determined by the decision-making rule that is used to aggregate conflicting interests into a single budget and they can affect spending bias.

This paper analyses a model in which the minister of finance internalizes the common pool budget's externality. From an institutional point of view, this assumption is realistic because he takes in account the budget equilibrium. Formally, this is reflected in the assumption that the minister of finance maximizes *à la* Stackelberg his utility function. In Stackelberg equilibrium, leader's expenditure choice is grater than in Cournot-Nash result, while the deficit bias is lower due to agenda setting power over spending ministers.

**Keywords** Common pool · Deficit bias · Cournot-Nash · Stackelberg

## 1 Introduction

The rationale for fiscal rules and institutions has been explained by the existence of deficit and spending biases that arise due to political fragmentation within government or between governments that alternate in office. The basic argument is that fragmented decision making increases the perspective on concentrated benefits of fiscal decisions for specific groups or during a specific period of time, while dispersing the costs in the form of general taxation over other groups in society or in time. Recent research shows that the origins of political fragmentation may go beyond political decision-making within the government itself. Persson and Tabellini [2] argue that the degree of political fragmentation within the government is related to the electoral rules in place. Political fragmentation within governments and between

G. Bimonte (✉)
Department of Economics and Statistics, University of Salerno, Fisciano (SA), Italy
e-mail: gbimonte@unisa.it

P. Spennati
Department of General State Account, Ministry of Economics and Finance, Rome, Italy
e-mail: pietro.spennati@tesoro.it

C. Perna, M. Sibillo (eds.), *Mathematical and Statistical Methods for Actuarial Sciences and Finance*, DOI 10.1007/978-3-319-05014-0_7,

governments are generally seen as the principal political sources of fiscal biases. The more fragmented is the system of budgetary decision-making, the weaker are the incentives for each participant to internalize the full tax burden of its spending bids so that a suboptimal level of spending results. Deficit and spending biases arise due to the political nature of fiscal decision-making. As a result, the incentives for biased policies could be removed or softened by taking (part of) the decision-making authority out of the political arena.

## 2 The Common Pool Problem

In common pool models fiscal outcomes are determined by the degree of political fragmentation and by the decision-making rule that is used to aggregate conflicting interests into a single budget. The rules according through which the budget is prepared, approved and carried out—in short the fiscal institutions—may therefore act to counteract political biases that are rooted in political fragmentation. In [3] the common pool problem may manifest itself during different phases of the budgetary process. When the budget is drafted within the cabinet, biases may arise due to the fact that spending ministers may recognize the full benefits of their own specific spending proposals, but fail to internalize the costs for the tax-paying population at large. Biases may again show up during the implementation phase of the budget, in the way policy reacts to unforeseen events and the way supplementary budgets are drafted, decided upon and implemented. von Hagen and Harden [4] consider a government consisting of $n$ spending ministers. The budget allocates public funds, raised through distorting taxation, to spending ministers, each of them pursuing its policy target. Collectively, the cabinet would wish to minimize the divergence between policy targets and actually allocated funds and, at the same time, to minimize the excess burden of taxation. In this model agents, having the same utility function subjected to the same budget constraint, interact simultaneously. The Cournot-Nash equilibrium shows that the bargaining between spending ministries reduces the spending bias due to the externality problem. The common pool problem arises from the fact that each spending minister takes into account only a share of that excess burden: the portion that falls on his constituency. From this premise, the budget realized by the cabinet is going to depend critically on the decision-making procedure. If the procedure entails collecting each minister's bid and taking a vote on the resulting budget, than the final budget will exhibit a spending bias.

In order to reduce the spending and deficit bias arising from the coordination problem in the budget process, we model a delegation of authority to a "fiscal entrepreneur" (the finance minister) with the aim to set binding limits on expenditure allocations collectively negotiated at the beginning of the budgeting process. The larger the finance minister's agenda-setting power, the closer the deficit comes to the collectively optimal outcome. Under this approach, the multilateral nature of the negotiations on fiscal targets implicitly forces all participants to consider the full cost in terms of tax burden associated with additional spending.

In [1] is modeled a weighted utility function in order measure the finance minister's power as an agenda setter. In this way they try to neutralize the incentive that single spending ministers will have to defect from the approved budget. Using a Nash-bargaining solution they show the larger the finance minister's agenda-setting power, the closer the deficit comes to the collectively optimal outcome.

## 3 Model and Results

We consider a two-period model of budgeting in a cabinet government. Consider a government consisting of $i = 1, \ldots, n+1$ spending ministers. Government expenditures consist of transfer $x_i$ to groups $i$ in society. Revenues consist of taxes levied on all groups of society and borrowing. In the first period borrowing must be repaid with interest in the second period. we assume that government can borrow or lend at a fixed real interest rate, $r$. We assume that government receives in the second period an amount $\tau_2$ of nontax revenue. The resulting intertemporal budget constraint involves a trade-off between the benefit from paying out more transfers in the first period and the cost of taxation in second period. The intertemporal utility function of each spending ministers is:

$$U(x_{t,i}) = -\frac{1}{2}\sum_{t=1}^{2} \delta^{t-1}\left[x_{t,i} - x^{\star}_{t,i}\right]^2 - m_i \Gamma(T)$$

with $i = 1, \ldots, n+1$, $\delta$ is the discount rate, $0 < \delta < 1$, $x_{t,i}$ is the level of spending allocated to minister $i$ and $x^{\star}_{t,i}$ the ideal level of spending from perspective of a single spending minister. We assume that $x^{\star}_i = x^{\star}_{1,i} = x^{\star}_{2,i}$. Each $m_i$ denotes the share of the excess burden from taxation falling on the minister $i$'s constituency, with $m_i < 1$ and for simplicity $m_i = 1/n$. The excess burden of taxation, i.e. the cost of taxation, is

$$\Gamma(T) = \frac{1}{2}\theta T^2.$$

The intertemporal spending minister's budget constraint over the two periods is

$$T = rB_1 + B_2 - \tau_2$$

where $B_t = \sum_{i=1}^{n+1} x_{t,i}$ for each period $t = 1, 2$. We first consider the case where all the spending ministers maximize their individual utility function subject to the intertemporal budget constraint, taking the other ministers' bids as given. The optimal levels for each individual spending ministers from the Cournot-Nash equilibrium are:

$$\begin{aligned}\widehat{x}_{1,i} &= x^{\star} - m_i \delta\theta r \widehat{T}\\ \widehat{x}_{2,i} &= x^{\star} - m_i \theta \widehat{T}.\end{aligned} \tag{1}$$

In the main literature the strong finance minister is modeled as a social planner maximizing a weighted intertemporal utility function. In this work, we consider the finance minister precommit fiscal policy and observes the $n$ spending ministers' optimal choices. This means that Finance Minister acts as a Stackelberg leader and the $n$ spending ministers as followers. The optimal choice of spending ministers remain the same in the previous model where they play Nash with each other. Instead, the finance minister's optimal choice is

$$\begin{aligned}\widetilde{x}_{1,1} &= x_1^\star - m_1\delta\theta r\widetilde{T}\left[1-\sum_{i=2}^{n+1} m_i\theta\left(1+\delta r^2\right)\right] \\ \widetilde{x}_{2,1} &= x_2^\star - m_1\theta\widetilde{T}\left[1-\sum_{i=2}^{n+1} m_i\theta\left(1+\delta r^2\right)\right].\end{aligned} \tag{2}$$

## 4 Conclusion

In this paper we assume that the minister of finance internalizes the common pool externality of the budget. From an institutional point of view, this assumption is realistic because he takes in account the budget equilibrium. Formally, this is reflected in the assumption that the minister of finance maximizes first (as a leader) his utility function. The finance minister's maximizing problem capture the different objective with respect to spending ministers: Finance Minister is benevolent i.e. internalizes spending bias. Comparing (1) and (2) reveals the nature of common pool problem. While main literature takes into account the role played by the portion $m_i$ of the cost of taxation when spending ministers making their budget bids, our result is related to the leader role of the finance minister, modelled *à la* Stackelberg. As finance minister's agenda setting power, his choices maximize utility subject to the condition that the proposal must be accepted by the spending ministers. In Stackelberg equilibrium, leader's expenditure choice is grater than in Cournot-Nash result, while the deficit bias is lower due to agenda setting power over spending ministers.

## References

1. Hallerberg, M., von Hagen, J.: Electoral institutions, cabinet negotiations, and budget deficits. In: Poterba, J., von Hagen, J. (eds.) Fiscal Institutions and Fiscal Performance. University of Chicago Press, Chicago (1999)
2. Persson, T., Tabellini, G.: Constitutional rules and fiscal policy outcomes. Am. Econ. Rev. **94**, 25–45 (2004)
3. von Hagen, J.: Budgeting procedures and fiscal performance in the European Communities. European Commission Economic Papers, 96, European Commission (1992)
4. von Hagen, J., Harden, I.J.: Budget processes and commitment to fiscal discipline. Eur. Econ. Rev. **39**, 771–779 (1995)

# Evaluating Correlations in European Government Bond Spreads

**Simona Boffelli and Giovanni Urga**

**Abstract** We propose a DCC-MIDAS model to estimate high- and low-frequency correlations in the 10-year government bond spreads. The high-frequency component, reflecting financial market conditions, is evaluated at 15-minute frequency, while the low-frequency one, fixed through a month, depends on country specific macroeconomic fundamentals. Although macroeconomic factors contribute in explaining volatilities and correlations, the increasing correlation in spreads during the pick of the sovereign debt crisis cannot be completely ascribed to macroeconomic factors.

**Keywords** DCC-MIDAS · Sovereign crisis

## 1 Introduction and Methodology

Since the introduction in 1999 of the Euro, the remarkable compression of sovereign risk premium differentials was considered a hallmark of successful financial integration in the Euro area. With the explosion of the sovereign debt crisis, government bond spreads started to diverge substantially as a sign of a regained ability of financial markets to careful monitor the fiscal performance of member states. Anyway, whether the yields were driven by countries fundamentals or rather by other factors, e.g. a regime shift in market pricing of government credit risk, was part of recent economic debate.

This paper is aimed at addressing that empirical question by extending the MIxed Data Sampling (MIDAS) framework which allows linking financial market data, sampled at high-frequency, with macroeconomic data recorded at lower frequency. In particular, we extend DCC-MIDAS in [1] based upon a pure time series approach

S. Boffelli (✉) · G. Urga
Università degli Studi di Bergamo, via dei Caniana 2, 24127 Bergamo, Italy
e-mail: simona.boffelli@unibg.it

G. Urga
Cass Business School, 106 Bunhill Row, London EC1Y 8TZ, UK
e-mail: g.urga@city.ac.uk

C. Perna, M. Sibillo (eds.), *Mathematical and Statistical Methods for Actuarial Sciences and Finance*, DOI 10.1007/978-3-319-05014-0_8,

by allowing the low-frequency component to be driven by country macroeconomic fundamentals. In details, we assess whether employment, industrial production and economic sentiment, recorded at monthly frequency, concur in explaining the pattern of 10-year government bonds spreads of Belgium, France, Italy, Spain and the Netherlands with respect to Germany measured on a 15-minute time scale over the period 1st June 2007–31st May 2012. Data were provided by Morningstar.

Let us consider a $(M \times 1)$ vector of returns for the $i$-th subinterval belonging to month $\tau$, $r_{\tau,i} = [r^1_{\tau,i}, \ldots, r^M_{\tau,i}]'$ distributed as a multivariate normal variable with mean vector $\mu$ and variance covariance matrix $H_{\tau,i}$ of order $(M \times M)$. Following the classical DCC model, the variance-covariance matrix $H_{\tau,i}$ can be decomposed as $D_{\tau,i} R_{\tau,i} D_{\tau,i}$ with $D_{\tau,i}$ diagonal matrix of volatilities and $R_{\tau,i}$ conditional correlation matrix. Volatilities are obtained by GARCH-MIDAS in [2], where the overall volatility can be decomposed into two parts, one pertaining to short term fluctuations, $g_{\tau,i}$ and the other to a long-run secular component, $\psi_\tau$:

$$r_{\tau,i} = \mu + \sqrt{\psi_\tau g_{\tau,i}} \varepsilon_{\tau,i}$$

where $\varepsilon_{\tau,i} | \Phi_{\tau,i-1} \sim N(0,1)$ with $\Phi_{\tau,i-1}$ information set available up to $(\tau, i-1)$.

The volatility dynamics of $g_{\tau,i}$ is modeled as a GARCH(1,1) process:

$$g_{\tau,i} = (1 - \alpha - \beta) + \alpha \frac{\varepsilon^2_{\tau,i-1}}{\psi_\tau} + \beta g_{\tau,i-1}$$

while the low-frequency component $\psi_\tau$ depends on macroeconomic variables:

$$\log \psi_\tau = m + \sum_{s=1}^{S} \vartheta^{s,l} \sum_{u=1}^{U} \varphi_u^{s,l}(\omega) X^{s,l}_{\tau-u} + \sum_{s=1}^{S} \vartheta^{s,v} \sum_{u=1}^{U} \varphi_u^{s,v}(\omega) X^{s,v}_{\tau-u}$$

where $X^{s,l}_{\tau-u} = abs(\frac{Y^{s,l}_{\tau-u}}{Y^{s,l}_{\tau 0}} - \frac{Y^{s,l,DE}_{\tau-u}}{Y^{s,l,DE}_{\tau 0}})$; $Y^{s,l}_\tau$ level $(l)$ of macroeconomic variable $s$ at month $\tau$; $Y^{s,l,DE}_\tau$ the same macrovariable $s$ for Germany, acting as benchmark country; $X^{s,v}_{\tau-u} = abs(Y^{s,v}_{\tau-u} - Y^{s,v,DE}_{\tau-u})$ where $Y^{s,v}_\tau$ volatility $(v)$ of macrovariable $s$; $\varphi_u(\omega)$ beta weights and $U$ maximum lag for macrovariable $s$, with $s = 1, \ldots, S$ with $S$ representing the total number of macroeconomic variables.

Once univariate volatilities are estimated, the main focus is on the correlation dynamics. In [1] it is shown that the high-frequency correlations obey a standard DCC scheme but with the intercept being a slowly moving process. Based on the DCC framework, the elements $\rho^{kj}_{\tau,i}$ of the conditional correlation matrix $R_{\tau,i}$, with $k, j = 1, \ldots, M$, are computed as:

$$\rho^{kj}_{\tau,i} = \frac{q^{kj}_{\tau,i}}{\sqrt{q^{kk}_{\tau,i}}\sqrt{q^{jj}_{\tau,i}}}$$

whose elements $q^{kj}_{\tau,i}$ are modeled by:

$$q^{kj}_{\tau,i} = \overline{\rho}^{kj}_\tau (1 - a - b) + a \xi^k_{\tau,i-1} \xi^j_{\tau,i-1} + b q^{kj}_{\tau,i-1}$$

where $\xi_{\tau,i}$ standardized residuals. The long-run correlation $\overline{\rho}_{\tau}^{kj}$ is obtained from the Fisher-z transformation of:

$$\gamma_{\tau}^{kj} = m^{kj} + \sum_{s=1}^{S} \vartheta^{s,l} \sum_{u=1}^{U} \varphi_{u}^{s,l}(\omega) \left| \Delta Y_{\tau-u}^{k;s,l} - \Delta Y_{\tau-u}^{j;s,l} \right|$$
$$+ \sum_{s=1}^{S} \vartheta^{s,v} \sum_{u=1}^{U} \varphi_{u}^{s,v}(\omega) \left( \Delta Y_{\tau-u}^{k;s,v} - \Delta Y_{\tau-u}^{j;s,v} \right)$$

where $\Delta Y_{\tau}^{k;s,l} = 100 \times [\ln(Y_{\tau}^{k;s,l}) - \ln(Y_{\tau-1}^{k;s,l})]$ and $|\Delta Y_{\tau}^{k;s,l} - \Delta Y_{\tau}^{j;s,l}|$ measure of the absolute distance in the rate of change for macrovariable $s$ during the period $(\tau, \tau - 1)$ between country $k$ and country $j$. $Y_{\tau}^{k;s,v}$ volatility of changes for macroeconomic fundamental $s$ for country $k$.

## 2 Empirical Results

Overall, the macroeconomic variables are found to be statistically relevant in explaining the volatility of European sovereign spreads, with the most important driver being the absolute difference between each country industrial production with respect to Germany: an increase of that difference determines a correspondent increase in volatility for all the countries but the Netherlands. As far as the economic sentiment is concerned, an increase in the absolute difference with respect to Germany implies a higher volatility for 3 out of 5 countries, Belgium, Italy and Spain, while employment is statistically significant just for the Netherlands. The differences in countries fundamental volatilities instead do not contribute in explaining volatilities of government bond spreads.

As per correlations, the macroeconomic variables turned out to be statistically significant drivers as well. Starting from the levels, there exists a negative relationship between correlations and the absolute differences in the rate of change of employment in 6 out of 10 pairs of countries, of industrial production in 4 out of 10 and of economic sentiment in 7 out of 10. Therefore, our results support the existence of a negative dependence between the correlation of two countries and the absolute difference in their macroeconomic fundamentals: as two countries get more similar in terms of their macroeconomic fundamentals, their respective government bond spreads start to move closer and closer. Focusing now on the absolute difference in the volatility of the rate of change of fundamentals, our results support the existence of a negative relationship with correlations for 4 out of 5 pairs of countries for which the estimates are statistically significant when the employment is taken into account, in 6 out of 6 for industrial production and in 5 out of 6 for economic sentiment.

Therefore, not only convergence in rates of change of macroeconomic variables determines an increase in correlation but the volatility of the rate of change too explains correlations in the same direction: as two countries get more similar in terms of volatilities of their fundamentals, their government bond spreads get even more correlated.

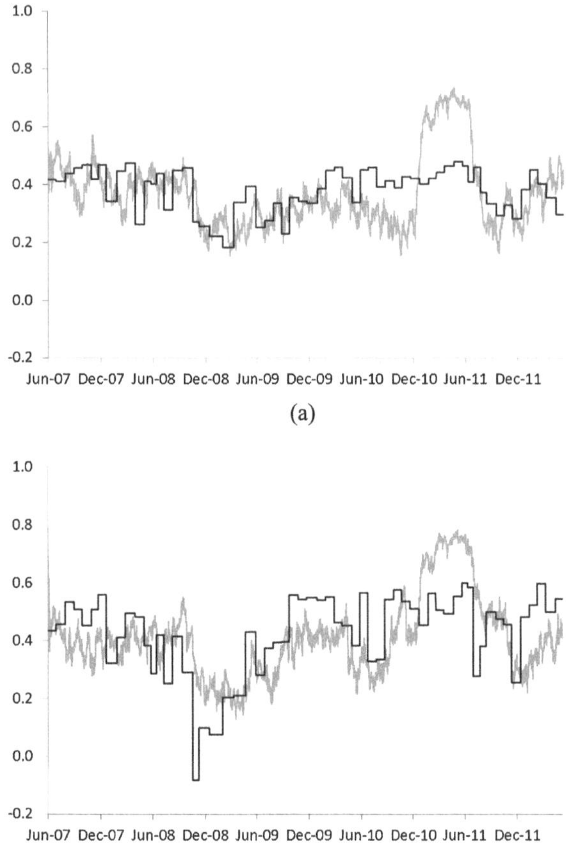


**Fig. 1** In *grey* is the high-frequency and in *black* the low-frequency component of correlation. (**a**) IT-FR; (**b**) IT-ES

In Fig. 1, we report the pattern of correlations between Italy and France and Italy and Spain (for Belgium and the Netherlands a similar pattern was identified).

First, we note the failure of the long run component driven by macroeconomic fundamentals (in black) in picking-up the break in correlations in financial markets (in grey) during the period December 2010–July 2011. This result sheds light in identifying the possible sources underlying the increasing systemic risk: the substantial break in correlations in government bond spreads, despite no change in correlations between countries fundamentals, shows that the increase in risk originated from financial markets rather than from shocks coming from the real side of economy. Second, the sharp increase in correlations is most likely due to a change in market sentiment as markets during crisis periods become more volatile and investment activities myopic. During the recent sovereign crisis this attitude was translated in a severe penalization of peripheral European countries in favour of Germany considered a "safe heaven".

## References

1. Colacito, R., Engle, R.F., Ghysels, E.: A component model for dynamic correlations. J. Econom. **164**, 45–59 (2011)
2. Engle, R.F., Ghysels, E., Sohn, B.: Stock market volatility and macroeconomic fundamentals. Rev. Econ. Stat. **95**, 776–797 (2013)

# Probability of Default: A Modern Calibration Approach

**Stefano Bonini and Giuliana Caivano**

**Abstract** An extensive academic and practitioner's literature exists on rating models development with well-structured statistical methods, however these models do not estimate PDs aligned with the economic scenario, then it is necessary a calibration. During the last years the effect of not well calibrated models has been observed on the credit market: actually they show a high level of procyclicality that let them loss market credibility and banking usability. The aim of this paper is to show a modern structured calibration approach, based on Bayesian techniques, taking into consideration specific economic factors. The calibration approach has been applied on real data of a Corporate portfolio of a top tier European Bank and a new calibration test, adjusted by the economic cycle, has been performed.

**Keywords** Rating models · Credit risk modeling · Bayesian econometric methods · Economic cycle

## 1 Introduction

Today, even because of the financial crises, banks need more and more reliable and usable risk management tools. Moreover within Basel2 and Basel 3 Accord the estimation of Probability of Default (PD) plays a key role for an efficient allocation of capital, pricing, client sanctioning, credit monitoring, and finally regulatory compliance.

A typical feature of PD models across countries is that they are often based on individual characteristics of clients or they use some information related to the type of specific products, but no information are commonly used for taking into account macroeconomic variables.

Typically the only way to align rating models with the economic scenario (as in [1] and [3]) is to apply a sort of "addendum" to the model itself: that's what is com-

S. Bonini (✉) · G. Caivano
University of Rome "Tor Vergata", via Columbia 2, 00133 Rome, Italy
e-mail: bonini.stefano@gmail.com

G. Caivano
e-mail: giuliana.caivano@gmail.com

C. Perna, M. Sibillo (eds.), *Mathematical and Statistical Methods for Actuarial Sciences and Finance*, DOI 10.1007/978-3-319-05014-0_9,

monly known as calibration. When speaking about calibration, it is a best practice among banks to refer to Through-the-Cycle (TTC) PDs as forecasts aligned with the average long run historical default rate typically used for capital requirement calculation. In this context the use of models that reflect the average long run historical default rate is required in order to stabilize capital ratios and optimize capital management policies: in positive economic conditions a run up of capital should be kept for negative economic phases. Given the relevance of final PDs in reflecting the economic conditions this paper proposes a modern structured calibration approach using Bayesian techniques for the estimation of average long run historical default rate (so called Central Tendency, CT). The added value of this work is related to the relevance itself of the topic in the last year: given the current context of crisis, it is very important for banks to forecast PDs that are able to ensure stable capital ratios. The methodology here proposed is easy to understand and also applicable to each kind of portfolio, as demonstrated by the application on real data of a Corporate portfolio of a top tier European Bank. In addition, the paper proposes a new binomial calibration test using PDs adjusted by the economic cycle.

## 2 Calibration Framework: Model Development and Validation Tests

As above mentioned, the goal of calibration is to assign implicit PD's to each grade or score defined by the rating model: in this way it is possible to define a consistent long-term PD based on the underlying scores defined by the statistical model.

In this context the Central Tendency has a key role, that is—among researchers and practitioners—quite often defined as the average of historical defaults. Here a new methodology is presented, in which the impact of a cyclical downturn in the future is embedded. For this scope, it is quite common [4] the use of macroeconomic variables such as GDP and their correlation with corporate defaults of each country in order to define a cut-off between positive and negative cycles [2]. Here we then define the correlation between quarterly Italian GDP, Italian corporate default rates, European corporate default rate and bank corporate portfolio default rate in order to forecast the future default and define the true picture of the portfolio in the current as well as in the future scenarios, even under stress hypotheses.

We here define the calibration function by solving an optimization problem with one objective function subject to the constraint that Central Tendency has to be equal to the implicit probability of default derived from calibration, as in (1):

$$\sum_{i=1}^{\#rating} PD_t^{Estimated} \times \%Pop_i = CT \tag{1}$$

where $\%Pop_i$ represents the percentage of observations in the bucket $i$ after proportioning.

The objective (target) calibration function has been defined according to (2):

$$CF = \sum_{i=1}^{n} \left[ \ln\left( \frac{1 - PD_i^{Estimated}}{PD_i^{Estimated}} \right) - \ln\left( \frac{1 - DR_i}{DR_i} \right) \right]^2 \quad (2)$$

where:

- $i' =$ score bucket $i$;
- $n =$ number of buckets;
- $RDF_i =$ observed default rate post re proportioning in the bucket $i$;
- $PD_i^{estimated}$ is the PD calculated as logistic transformation of scores of statistical model.

According to the Bayesian approach [5] the default rates of each bucket have been proportioned in order to align them with the Central Tendency as shown in (3).

$$Adj.DR_{bucket} = \frac{DR_{bucket} \frac{CT}{DR_{sample}}}{DR_{bucket} \frac{CT}{DR_{sample}} + (1 - DR_{bucket}) \frac{1-CT}{1-DR_{sample}}} \quad (3)$$

A model calibrated according to cyclical effect generate PDs that could be slightly different from the observed default rates during the most recent period (e.g. the last year of observation) in this case the standard calibration binomial tests (based on the null hypothesis $H_0$ that the PD of each rating grade is correctly estimated)[1] could "wrongly" fail [6]. The authors propose then a new calibration test called "cycle adjusted" in which PDs are "adjusted for the economic cycle trend, as defined in (4):

$$PD_{1,t}^{adj} = N\left[ \frac{N^{-1}(PD_1) - \sqrt{AC} \times N^{-1}(\Delta Y^t)}{\sqrt{1 - AC}} \right] \quad (4)$$

where:

- $PD_i =$ Average PD of each rating class;
- $AC =$ Asset correlation of defaults;
- $\Delta Y =$ percentile of empirical GDP variation at time $t$.

The calibration function has been estimated on 7 years of historical data (2004–2011) of a Corporate portfolio of a top tier European Bank: the gradient ($\beta$) of curve has been considered as a constant, while the intercept ($\alpha$) has been changed in order to ensure that the average PD of the portfolio will be equal to the CT. In particular, taking into account the function constraint (average PD equal to CT), the final values of $\alpha$ ($-50.02$) and $\beta$ ($-115.97$) have been found as the values that could minimize the difference between estimated PD and proportioned default rate, as in (3).

[1] $k^* = \phi^{-1}(q)\sqrt{nPD(1 - PD)} + nPD$
where:
$q =$ is the confidence level of test;
$PD =$: is the theoretical PD of each rating grade;
$n =$ number of observations.

This function has been applied on a portfolio of Corporate loans existing in 2012, thus only the value of intercept has been changed in order to align the average PDs to CT, obtaining a new value of $\alpha$ equal to $-32.20$.

Finally, an adjusted binomial test has been performed, in which the adjusted PDs have been used (as defined in (4)) for assessing the goodness of the overall model after the calibration process.

## 3 Conclusion

In this paper a new approach of calibration has been proposed for aligning "traditional" rating models to the economic cycle by avoiding, at the same time, the pro-cyclical effect of using long run estimates. In particular, a new approach for the definition of Central Tendency has been proposed by forecasting the historical default rates on a "long run" period (10 years) to be used for fitting the calibration curve of the application portfolio represented by Corporate exposures of a top tier European bank between 2004–2012. The authors have also performed a binomial test "adjusted" for the state of the economy, in order to avoid the underestimation of PDs during the last recent years: the results of the test show that the estimated PDs are conservative.

## References

1. Engelmann, B., Porath, D.: Do not forget the economy when estimating default probabilities. Willmott Mag. (2012)
2. Iqbal, N., Ali, A.: Estimation of Probability of Defaults (PD) for low default portfolios: an actuarial approach. In: ERM Symposium (2012)
3. Kiff, J., Kisser, M., Schumacher, L.: Rating through-the-cycle: what does the concept imply for rating stability and accuracy? IMF Working Paper (2013)
4. Konrad, M.P.: The Calibration of Rating Models. Tectum, Marburg (2012)
5. Tasche, D.: The art of PD curve calibration. J. Risk Manag. (2013)
6. van der Burgt, M.: Calibrating low-default portfolios, using the cumulative accuracy profile. J. Risk Model Valid. (2008)

# Development of a LGD Model Basel2 Compliant: A Case Study

**Stefano Bonini and Giuliana Caivano**

**Abstract** The Basel2 Accord allows banks to calculate their capital requirements using Advanced Internal Ratings Based Approach (AIRBA) based on the estimation of three credit risk parameters—Probability of Default (PD), Exposure at Default (EAD) and Loss Given Default (LGD). While on PD models an extensive academic and practitioner's literature exists, LGD studies are in a less advance status because of the lack of data on recoveries of commercial loans: the existing literature on LGD is for the most part related to Corporate Bonds. In this paper a case study on a real Basel 2 compliant model has been developed starting from a *workout approach* and stressing on estimation of the discount rate as main component of Economic LGD but also on the definition of the final multivariate regressive model.

**Keywords** Loss given default · Basel2 · Credit risk modeling · Quantitative finance

## 1 Introduction

In the last years the biggest European Banking Groups started to assess the possibility of adopting the Advanced Internal Rating Based Approach (AIRBA) under Basel2, in order to save capital thanks also to the possibility of a larger use of Credit Risk mitigators with respect to the Standardized Approach. The AIRBA framework requires banks to develop statistical models for estimating probability of default (PD), Loss Given Default (LGD) and Exposure at Default (EAD). In particular the LGD, that is defined as credit loss when extreme events occur influencing the obligor ability to repay debts, has a high relevance into credit and recovery process because of its direct impact on capital savings.

The New Basel2 Accord, which has been implemented throughout the banking world starting from 1 January 2007, made a significant difference to the use of

S. Bonini · G. Caivano (✉)
University of Rome "Tor Vergata", via Columbia 2, 00133 Rome, Italy
e-mail: giuliana.caivano@gmail.com

S. Bonini
e-mail: bonini.stefano@gmail.com

C. Perna, M. Sibillo (eds.), *Mathematical and Statistical Methods for Actuarial Sciences and Finance*, DOI 10.1007/978-3-319-05014-0_10,

modeling within financial organizations, by highlighting the relevant role of Loss Given Default (LGD) modeling [2].

While on PD models an extensive academic and practitioner's literature exists, LGD studies are in a less advance status because of the lack of data on recoveries and differences in recovery process among commercial banks. The existing literature on LGD is for the most part related to Corporate Bonds given the public availability of data (as in [1] and [6]) and in most cases the existing papers only try to test different statistical approaches in order to identify the most predictive variables/methods of recovery rates (as in [8, 9], and [3]).

The aim of this paper is to show the results of a case study on a real Basel2 compliant LGD model starting from a *workout approach* and stressing on the estimation of the discount rate as main component of Economic LGD, but also on the definition of the final multivariate regressive model for the LGD estimation. The model has been developed on 10 years of historical real data of Corporate and Retail portfolio of an Italian commercial bank among the fifteen Italian Banks that will be supervised by ECB. This paper adds a real value to the existing literature because it follows all the Basel2 requirements and it is linked to the Italian Banking context. Italy, unlike the rest of Europe, can be considered a more general and complicate case of LGD computation because of specific recovery process and more than one default status (doubtful loans, past-dues, charge-offs). This paper, finally, contributes to the understanding of the Italian recovery process in order to define an even more common European framework looking forward the ECB Banking Supervision.

## 2 LGD Model Development

The LGD model proposed in this paper has been developed on a sample of around 25,000 charge-offs loans (opened and closed between January 2000 and December 2012) belonging to Corporate and Retail (40 % and 60 % respectively) clients with different types of products: 30 % mortgages, 65 % checking accounts, and 5 % others.

The best practice on European Banks, in particular on Corporate and Retail Portfolios, is to use a *workout approach* [4]. It is based on economic notion of loss and consists in the calculation of empirical loss rates through the observation of historical cash flows (discounted for taking into account the volatility embedded in time) occurring on each charge-off until the end of recovery process, according to (1) and the definitions in Table 1:

$$LGD_c = 1 - RR = 1 - \frac{\sum Rec_i \delta_i^t - \sum A_i \delta_i^t - Costs}{EAD} \quad (1)$$

For the calculation of economic LGD, and in particular the definition of discount rate, the Capital Asset Pricing Model (CAPM) has been used based on the theoret-

**Table 1** List of factors for LGD calculation on charge-off positions

| Parameter | Description |
|---|---|
| $LGD_C$ | LGD estimated on charge-offs positions |
| $RR$ | Recovery rate on charge-offs |
| $REC_i$ | Recovery flow at date $i$ |
| $A_i$ | Increase flow at date $i$ |
| $Costs$ | Costs of litigation |
| $EAD$ | Exposure of default at charge off opening date |
| $i$ | Date in which each flow has been registered |
| $t$ | Charge—off opening date |
| $\delta_i^t$ | Discount rate of each flow between date $i$ and date $t$ |

**Table 2** Discount rate definition: main parameters

| CAPM parameter | Description |
|---|---|
| Market volatility—$\sigma_M$ | Standard deviation of logarithmic returns of stock market |
| Asset volatility—$\sigma_i$ | Standard deviation of logarithmic returns of the ratio between annual recoveries and total exposure at default |
| Correlation $R_i$ | Basel2 correlation framework for capital requirements |
| Market Risk Premium | Set at 5.6 % according to [5] |
| Risk Free component | Defined from a linear interpolation of interest rate curves |

ical framework proposed by [7] in (2) and according to the meaning of parameters provided in Table 2:

$$r_t = r_t^{RF} + \beta_{i,M} \times MRP = r_t^{RF} + \frac{\sigma_i \rho_{i,M}}{\sigma_M} \times MRP = r_t^{RF} + \frac{\sigma_i R_i}{\sigma_M} \times MRP \tag{2}$$

The methodology applied is based on the hypothesis that when a corporate defaults, its credit contract can be considered as a potential investment contract: thus the discount rate must reflect the cost-opportunity of this investment.

The risk premium[1] for retail clients as been set to 0.48 % for Mortgages and 0.45 % for Other Products, instead for corporate clients it has been set to 0.50 %.

A final average value of Economic LGD of 50 % has been obtained on development sample, and the Economic LGD has been used as target variable of the multivariate regression model, consisting in a linear regression based on Ordinary Least Square (OLS).[2]

In Table 3 the final model is shown with the main drivers of recovery identified.

[1]The value of risk premium has been differentiated according to the correlation values of capital requirements calculation formula of Basel2 framework.

[2]SAS PROC GLM procedure has been used.

**Table 3** Final Model

| Variables | Values |
|---|---|
| Geographical area | South & Islands, Center, North West, and North East |
| Exposure at Default | $EAD \leq 1{,}500$, $1{,}500 > EAD \leq 2{,}500$, $2{,}500 > EAD \leq 7{,}500$, $7{,}500 > EAD \leq 15{,}000$. and $EAD > 15{,}000$ |
| Type of product | Residential Mortgages, Checking Accounts, and Other products |
| Type of segment | Retail, Corporate |
| Presence of Prs Guarantees | |
| Presence of Pledge | |

## 3 Conclusion

This paper has presented a case study of LGD in which, according to the requirements of Basel2, the model has been developed on 10 years of historical real data of Corporate and Retail portfolio of an Italian commercial bank among the fifteen Italian Banks that will be supervised by ECB. Giving a particular stress on the economic component of the model, the presented model highlights the determinant role of mitigators as recovery drivers, but also the geographical localization of loans, the loan and commercial segment. This paper adds a real value to the existing literature because it follows all the Basel2 requirements and is linked to the Italian Banking context. Italy, unlike the rest of Europe, can be considered a more general and complicate case of LGD computation because of specific recovery process and more than one default status (past-due, doubtful, and charge-off). Moreover it contributes to the understanding of the Italian recovery process in order to define an even more common European framework looking forward the ECB Banking Supervision.

## References

1. Altman, E., Brady, E., Resti, A., Sironi, A.: The link between default and recovery rates: theory, empirical evidence, and implications. J. Bus. (2005)
2. Banca d'Italia: Circolare n. 263. Titolo 2—Capitolo 1, ultimo aggiornamento (2013)
3. Bonini, S., Caivano, G.: Survival analysis approach in Basel2 credit risk management: modelling danger rates in loss given default parameter. J. Credit Risk (2013)
4. CEBS: GL10 guidelines on the implementation, validation and assessment of Advanced Measurement (AMA) and Internal Ratings Based (IRB) approaches (2006)
5. Fernandez, P.: Market Risk Premium Used in 82 Countries in 2012: A Survey with 7,192 Answers. IESE Business School Publication (2012)
6. Jacobs, M.: Measuring LGD on commercial loans: an 18-year internal study. RMA J. (2004)
7. Maclachlan, I.: Choosing the discount factor for estimating economic LGD. In: Recovery Risk: The Next Challenge in Credit Risk Management. Bloomberg Financial (2004)
8. Schuermann, T.: What do we know about loss given default? Working paper on Wharton Financial Institutions Center (2004)
9. Yashir, O., Yashir, Y.: Loss given default modeling: a comparative analysis. J. Risk Model Valid. (2012)

# Modelling the Latent Components of Personal Happiness

**Stefania Capecchi and Domenico Piccolo**

**Abstract** We discuss a class of statistical models able to measure the self-evaluation of happiness by means of a sample of respondents and investigate the ability of this proposal to enhance the different contribution of subjective, environmental and economic variables. The approach is based on a mixture model introduced for interpreting the ordered level of happiness as a combination of a real belief and a surrounding uncertainty: these unobserved components may be easily parameterized and immediately related to subjects' covariates. An empirical evidence is supported on data set derived by the Survey of Household Income and Wealth (SHIW) conducted by the Bank of Italy.

**Keywords** Happiness · Ordinal data · CUB models · SHIW data set

## 1 Introduction

According to dictionaries, happiness is a state of well-being characterized by emotions ranging from contentment to intense joy. Together with life and liberty, the pursuit of happiness has been considered an unalienable right by the US Declaration of Independence. Specifically, intangible goods are a relevant issue in the approach promoted by the Stiglitz Commission [7]. In recent years, several countries introduced new measures of global subjective happiness and noticeably in 1972 the Kingdom of Bhutan established the Gross National Happiness (GNH) measure as a multidimensional indicator of people well-being and satisfaction of life. In Italy, the ISTAT experience of BES is a remarkable one.

Happiness is a concept perceived in different ways by people and it is often used as synonym of flourishing quality of life. Scientists are engaged to derive origin and causes of happiness, and it is now a common evidence that it derives from a blend

S. Capecchi · D. Piccolo (✉)
Department of Political Sciences, University of Naples Federico II, via L. Rodinò 22, 80138 Naples, Italy
e-mail: domenico.piccolo@unina.it

S. Capecchi
e-mail: stefania.capecchi@unina.it

C. Perna, M. Sibillo (eds.), *Mathematical and Statistical Methods for Actuarial Sciences and Finance*, DOI 10.1007/978-3-319-05014-0_11, 

of internal and external pleasures, including monetary achievements. In fact, people living in more developed nations tend to be on average happier than those of less developed ones. Although the effect is not linear and diminishes with wealth, this circumstance explains why GDP and GNP have been used as measures of successful policies. On the other side, the well known Easterlin paradox [2] assesses that once wealth reaches a subsistence level, its effectiveness as a generator of well-being is greatly diminished. Thus, indicators and models to face with a fuzzy concept as happiness are both a need and a challenge.

This work relies on a statistical model elaborated for investigating ordinal responses and checks its effectiveness on the Survey of Household Income and Wealth (SHIW) conducted by the Bank of Italy. The scope is to emphasize the effective possibility to investigate the interpretation of the declared happiness as a function of subjective, contextual and economic covariates.

The paper is organized as follows: in the next section the data set is described and the model chosen for the statistical analysis is briefly motivated. In Sect. 3 the main results are presented.

## 2 Data and Model

Several scales may be considered to measure happiness by means of subjective evaluation. In a large and accurate data set as the 2010 edition of SHIW, responses are expressed on a Likert 10-point scale. It is sufficiently fine to allow adequate expression of personal opinions and our analysis is based on $n = 3816$ validated questionnaires.

Economists have been skeptical about subjective data but recent approaches based on the relationship between economic variables and components of well-being emphasize the importance of subjective evaluation data [7]. A further consideration derives from the common assumptions of economists who assume that people make decisions basically on expectations for unknown quantities and so to maximize expected utility. Indeed, when using ordinal data, the use of expectation has to be critically considered because several distributional shapes are admissible for a given expectation. Thus more comprehensive approaches should be pursued.

Specifically, we consider ordinal responses as a manifest piece of information which conveys both the latent variable to be measured and some intrinsic indecision. Then, we assume that the psychological expression of the degree of happiness is the composition of a real *perception* generated by a continuous random variable and a second component pertaining to an inherent *uncertainty*.

These considerations motivated the introduction of CUB models [1, 6] defined as a mixture distribution in which the rating $r$ is the realization of a random variable $R$ with probability mass which is a Combination of a (shifted) Binomial and a (discrete) Uniform random variable, that is:

$$Pr(R = r) = \pi \binom{m-1}{r-1} \xi^{m-r} (1-\xi)^{r-1} + (1-\pi)\frac{1}{m}, \quad r = 1, 2, \ldots, m.$$

These models are identifiable for a given $m > 3$ [3]. The uncertainty (measured by $1 - \pi$) and the perception (measured by $1 - \xi$) may be easily visualized in the parameter space (that is, the unit square); thus, it is immediate to see the role of these components. In our context, $(1 - \xi)$ may be interpreted as perception of happiness whereas $(1 - \pi)$ explains the degree of indecision of the respondent.

This class of models has been applied in many different fields and generalized in several directions [4, 5]; in the following, we limit ourselves to show the potential of the approach in its standard formulation.

## 3 Main Results

The selection of significant covariates is a relevant issue in the building of econometric models and it deserves specific attention when the response is an ordinal variable as the expressed rating of happiness. In this regard, we performed a stepwise strategy mainly based on the joint consideration of parameters significance and increase in log-likelihood measures. The omnibus CUB model we present is the final result of this sequential approach and expresses the probability of the ordinal response as the consequence of both perception and uncertainty related to significant covariates.

If we set $\text{logit}(z) = [1 + \exp(-z)]^{-1}$, then we may consider $\pi_i = \text{logit}(\boldsymbol{\beta}\boldsymbol{y}_i)$ and $\xi_i = \text{logit}(\boldsymbol{\gamma}\boldsymbol{w}_i)$, where $\boldsymbol{y}_i$ and $\boldsymbol{x}_i$ are the significant covariates of the $i$-th subject, for $i = 1, 2, \ldots, n$, selected to explain uncertainty and perception, respectively. Table 1 shows the estimates of the CUB model. All parameters are significant (standard errors in parenthesis) and results are effectively obtained by maximum likelihood methods (a program in R is freely available from Authors).

It turns out that some covariates are subjective (`age`, `married`, `confidence`), environmental (`South`) and economic (`familycond`, `wealth`). Notice that `confidence` is related to the general attitude of trust towards the others whereas `familycond` expresses if *household income is sufficient to see the family to make ends meet* and ranges from 1 (*with great difficulty*) to 6 (*very easily*). The estimated model shows that happiness is positively related with `married`, `familycond`, `confidence`, `wealth` whereas `age` and `South` have a negative effect; the effect of `age` is negative when people are aged more than 58.6 years. On the other side, uncertainty increases for `married` and `confidence` but decreases with `age`, `South`, `familycond` and `wealth`. It is also possible to show that `married`, `familycond` and `wealth` give the best explanation to the perception of happiness whereas `married` and `confidence` give the most relevant contribution to explain the uncertainty.

The proposed approach allows to depict expected profiles of responses given some values for covariates. As an instance, given approximately the same wealth, we compare in Fig. 1 the probability distributions of declared happiness of an unmarried resident in South and aged 30 with low `confidence` and `familycond` (dotted left-shifted) to the happiness of a resident elsewhere, married and aged 57 with high `confidence` and `familycond` (squared right-shifted). It is quite evident how happiness changes in a substantial way when the significant covariates are modified.

**Table 1** Estimated CUB model

| *Covariates* | $\hat{\beta}_k$ | $\hat{\gamma}_k$ |
|---|---|---|
| *Constant* | −2.653 *(0.928)* | 1.576 *(0.319)* |
| Age-dev | | 0.008 *(0.001)* |
| Married | −1.173 *(0.144)* | −0.521 *(0.047)* |
| South | 0.351 *(0.144)* | 0.231 *(0.047)* |
| Familycond | −0.204 *(0.063)* | −0.091 *(0.020)* |
| Confidence | 0.117 *(0.025)* | −0.054 *(0.012)* |
| Wealth-Log | 0.273 *(0.081)* | −0.119 *(0.028)* |

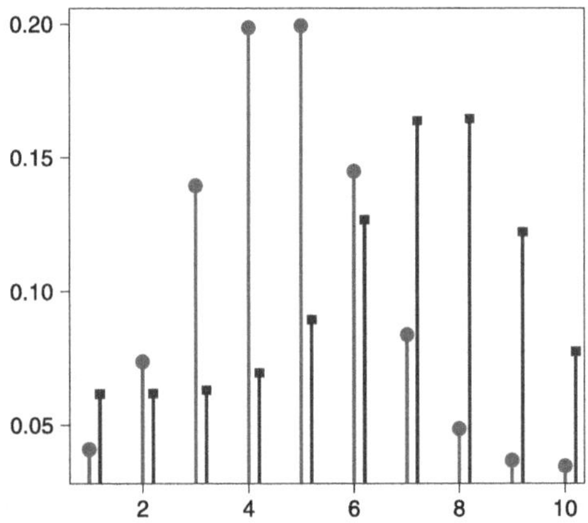


**Fig. 1** Different respondents profiles

**Acknowledgements** This work has been partly supported by projects FIRB 2012 "Mixture and latent variable model for causal inference and analysis of socio-economic data" (code RBFR12SHVV), University of Perugia, and STAR 2013, University of Naples Federico II.

## References

1. D'Elia, A., Piccolo, D.: A mixture model for preference data analysis. Comput. Stat. Data Anal. **49**, 917–934 (2005)
2. Easterlin, R.: Will raising the incomes of all increase the happiness of all? J. Econ. Behav. Organ. **27**, 35–48 (1995)
3. Iannario, M.: On the identifiability of a mixture model for ordinal data. Metron **LXVIII**, 87–94 (2010)
4. Iannario, M.: Modelling shelter choices in a class of mixture models for ordinal responses. Stat. Methods Appl. **21**, 1–22 (2012)
5. Iannario, M., Piccolo, D.: CUB models: statistical methods and empirical evidence. In: Kenett, R.S., Salini, S. (eds.) Modern Analysis of Customer Surveys: With Applications Using R, pp. 231–258. Wiley, Chichester (2012)
6. Piccolo, D.: On the moments of a mixture of uniform and shifted binomial random variables. Quad. Stat. **5**, 85–104 (2003)
7. Stiglitz, J.E., Sen, A., Fitoussi, J.P.: Report by the Commission on the Measurement of Economic Performance and Social Progress (2009)

# Measuring the Impact of Behavioural Choices on the Market Prices

**Massimiliano Caporin, Luca Corazzini, and Michele Costola**

**Abstract** We present a methodology to build a new sentiment index of market (ir)rationality. The proposed index, derived only on the basis of equity market prices, could be used to monitor the impact on behavioural-driven agent's choices. In this note, we discuss the main idea behind the proposed approach.

**Keywords** Investment decision · Behavioural agents · Mixture model · Behavioural expectations

## 1 Introduction

According to the traditional theory of finance, see [5], in taking their financial investment decisions, agents are assumed to be rational in that they make their choices by maximising a utility function that is consistent with the Expected Utility Theory (EUT). However, the rational hypothesis is inconsistent with several empirical puzzles, see [1], among others.

Those evidences lead the development of models including behavioral and psychological elements in the agents' decision process, namely, agents are not fully rational, [3]. One of the most successful approaches is the Prospect Theory [4], where the (risk) preferences of agents are described by a value function making them risk-seeker in the domain of losses and risk averse in the domain of gains.

Despite the several generalizations appeared in the financial literature, the Prospect Theory leaves important empirical issues unexplored. One of the most

M. Caporin (✉) · L. Corazzini
Department of Economics and Management Marco Fanno, University of Padova, via del Santo 33, 35123 Padova, Italy
e-mail: massimiliano.caporin@unipd.it

L. Corazzini
e-mail: luca.corazzini@unipd.it

M. Costola
Department of Economics, University Ca' Foscari Venezia, San Giobbe 873, 30121 Venezia, Italy
e-mail: michele.costola@unive.it

C. Perna, M. Sibillo (eds.), *Mathematical and Statistical Methods for Actuarial Sciences and Finance*, DOI 10.1007/978-3-319-05014-0_12,

relevant is associated with the measurement of the behavioural component of the financial market. In fact, we might assume that the market is composed by agents that are either rational (endowed with a risk averse utility function that is coherent with the standard EUT assumptions) or behavioural (with a S-shaped utility function). By means of this setting, we can investigate into the relative weights of the two categories and analyse how the behavioural component changes over time. Intuitively, in periods of recessions we might expect a divergence of investment decisions, with behavioural agents taking more risky investment decisions compared to rational ones.

In this note, we introduce a Bayesian mixture approach to estimate the relative weight of the behavioural component of the financial market. In a market that is populated by two types of non-strategic financial agents (rational and behavioural), the evolution of the prices reflects the interplay between their choices. We thus define the investment decision of agents as driven by a specific performance measure, which allows them to build ranks of assets and use those ranks to define their portfolio. Given their portfolio, agents are assumed to invest in the subset of the investment universe including the best performing assets. In such a framework, the financial market produces a mixture ranking that is built by conditioning the prior ordering of the rational, risk-averse agents on that produced by the behavioural category. The mixture depends on a weighting factor that expresses the relative weight of the behavioural category over the rational one: the higher the value of the weighting factor, the more the ranking of the financial market approaches the one produced by the behavioural agents.

In the following we introduce the model that depends on the weighting factor and describe its estimation approach.

## 2 Agents and Market Behaviour

As mentioned above, our framework postulates the existence of two types of agents that choose their optimal allocation and make their evaluations in terms of performance measures at the level of the single asset. Performance measures are related to the level of maximum expected utility provided by a given asset and are function of the moments of the risky assets returns distribution. The higher the performance measure, the higher the maximum expected utility that the investor attaches to the asset. The allocation choice of the agent is made by investing in a subset composed of the most performing assets—namely, those with the highest performance measures.

The first agent is *rational*, with a utility function consistent with the expected utility theory, namely a negative exponential utility. The second agent is *behavioural* and with the generalized *behavioural* utility function of [6]

$$U(W) = \begin{cases} 1_{+}(W - W_0) \times (W - W_0) - (\gamma_{+}/\alpha)(W - W_0)^{\alpha}, & \text{if } W \geq W_0, \\ -\lambda(1_{-}(W_0 - W) \times (W - W_0) + (\gamma_{-}/\beta)(W_0 - W)^{\beta}), & \text{if } W < W_0, \end{cases} \tag{1}$$

where: $1_+(\cdot)$ and $1_-(\cdot)$ are the indicator functions, while $\gamma_+$, $\gamma_-$, $\lambda > 0$, $\alpha > 0$ and $\beta > 0$ are real numbers. The agents allocate their wealth over a set of risky assets by analysing the expected utility of each single asset, investing only in the best performing. We are thus interested in the rankings of the performance measures that are produced by the *rational* and *behavioural* utility functions.

For the rational agent, we follow [7] and consider the Generalized Sharpe ratio (*GSR*), obtained by the numerical optimization of the expected utility,

$$GSR = \sqrt{-2\log(-E[U(\tilde{w})])}, \tag{2}$$

$$E[U(\tilde{w})] = max_a \int -e^{-\lambda a(x-r_f)} \hat{f}(x)dx \tag{3}$$

with $x$ being the return of a risky asset, $r_f$ the return of the risk-free asset, $\hat{f}(x)$ a Kernel estimate of the risky asset returns density, and $\lambda$ the risk aversion coefficient.

The behavioural agent ranks assets according to a different index, the $Z$-ratio, see [6],

$$Z_{\gamma_-,\gamma_+,\lambda,\beta} = \frac{E(x) - r_f - (1_-(W - W_0)\lambda - 1)LPM_1(x, r_f)}{\sqrt[\beta]{\gamma_+ UPM_\beta(x, r_f) + \lambda\gamma_- LPM_\beta(x, r_f)}}, \tag{4}$$

where *LPM* and *UPM* are the lower and upper partial moments.

Both agents allocate their wealth on the best performing assets. We assume that the allocation choice follows a naive rule. If the market includes $K$ assets, we might assume that the rational (behavioural) investor allocates his wealth across the $M \ll K$ assets with highest value of the *GSR* ($Z$-ratio) with an equally weighted strategy. Such a choice allows limiting the impact of the estimation error and is preferred over optimal weighting schemes, see [2].

While both types populate the market, one could question whether market fluctuations are more closely related to the choices of the rational rather than the behavioral agents—irrespective of their numerosity. Our objective is to determine the relevance or the impact of the behavioural choices in the movements of risky asset returns. We propose to recover such a measure in an indirect way by assuming that the observed market behaviour is a blend of choices made by the different choices made by the two types. Given this framework, we estimate the blending parameter(s) in such a way that the combination of choices is as closer as possible to the observed market fluctuations. We blend the choices of the two agents types within a Bayesian framework where one of the two agent's beliefs is considered a prior (the rational), while choice of the behavioural agents plays the role of additional conditioning information. The beliefs of the agents are given by their performance measures, the *GSR* and the $Z$-ratio.

We thus derive the posterior aggregate performance measure:

$$\mu_p = [(\tau\sigma^2)^{-1} + \omega^{-2}]^{-1}[(\tau\sigma^2)^{-1} GSR + \omega^{-2} Z_{\gamma_-\gamma_+,\lambda,\beta}] \tag{5}$$

where $\sigma^2$ is the variance of the *GSR*, $\omega^2$ is the variance of the *Z-ratio*, and $\tau$ is our main objective, the parameter driving the reliability of the prior information.

The higher the $\tau$ the more uncertain the prior rational information and, conversely, the more relevant the behavioural impact on aggregate rankings. The aggregate expected measure $\mu_p$ might be considered as the quantity used, at the market level, to order or rank assets. As a consequence, we determine the role of behavioural choices through the composite measure, by looking at the optimal allocation made by an agent which is deciding where to invest his wealth across a set of risky assets ordered according to (5). In this case, the allocations might be evaluated in terms of past performances, while the impact of behavioural beliefs is determined by estimating the optimal $\tau$ level within a specified criterion function. We take a simplified allocation choice and consider an equally weighted investment strategy. The optimal choice of $\tau$ is determined by maximizing the portfolio returns, that is

$$\max_{\tau} \quad f(\tau) = \frac{1}{m} \sum_{l=t-m+1}^{t} r_{p,l} \tag{6}$$

$$\text{s.t.} \quad r_{p,l} = \frac{1}{k} \sum_{j \in \mathcal{A}_t(\tau)} r_{j,l} \tag{7}$$

where the set $\mathcal{A}_t(\tau)$ contains the $M$ best assets (with highest aggregate posterior performance measure) across the $K$ included in the investment universe. The optimal value $\tau^*$ provides the maximum cumulated return obtained by an agent investing in a subset of the risky assets traded in the market and taking decisions blending rational and behavioural choices.

## References


1. Barberis, N., Thaler, R.: A survey of behavioral finance. In: Constantinides, G.M., Harris, M., Stulz, G.M. (eds.) Financial Markets and Asset Pricing. Handbook of the Economics of Finance, pp. 1053–1128. Elsevier, Amsterdam (2003)
2. DeMiguel, V., Garlappi, L., Uppal, R.: Optimal versus naive diversification: how inefficient is the $1/n$ portfolio strategy? Rev. Financ. Stud. **22**(5), 1915–1953 (2009)
3. Hommes, C.H.: Heterogeneous agent models in economics and finance. In: Tesfatsion, L., Judd, K.L. (eds.) Handbook of Computational Economics, vol. 2, pp. 1109–1186. Elsevier, Amsterdam (2006)
4. Kahneman, D., Tversky, A.: Prospect theory: an analysis of decision under risk. Econometrica **47**(2), 263–292 (1979)
5. LeRoy, S.F., Werner, J.: Principles of Financial Economics. Cambridge University Press, Cambridge (2000)
6. Zakamouline, V.: Portfolio performance evaluation with loss aversion. Quant. Finance 1–12 (2011)
7. Zakamouline, V., Koekebakker, S.: Portfolio performance evaluation with generalized Sharpe ratios: beyond the mean and variance. J. Bank. Finance **33**(7), 1242–1254 (2009)

# A Note on Natural Risk Statistics, OWA Operators and Generalized Gini Functions

**Marta Cardin**

**Abstract** The notion of risk measure arose from the problem of quantifying risk. The coherent risk measures and the insurance risk measures are risk measures that satisfy a set of axioms. In this note we consider a different approach to risk measurement and we study natural risk statistics that are based on data and that are characterized by a new set of axioms. In this paper we consider the relationship between risk measurement and aggregation theory.

**Keywords** Risk measure · Coherent risk measure · Natural risk statistic · Gini index

## 1 Introduction

During the last decades, researchers joined efforts to properly compare, quantify and manage risk. In this direction, risk measures constitute an important and widely studied tool. Different families of risk measures have been proposed in the literature. The paper that lays the foundations of the axiomatic approach in defining a risk measure is [4].

In 2007 Heyde et al. [6] introduced the natural risk statistics that are risk measures depending on data (see also [1]). The natural risk statistics are associated with a finite sample and satisfy a more general subadditivity assumption then that of classical coherent risk measures and are robust thus particularly suitable for external risk measurement. The aim of this note is to study and characterize some classes of natural risk statistics that are defined by ordered weighted averaging operators. These operators were introduced in aggregation theory by Yager and have been employed in a wide range of fields and only recently considered in risk measurement. Moreover we introduce in risk measurement the generalized Gini welfare functions that are traditionally studied as inequality indices.

M. Cardin (✉)
Department of Economics, University Cà Foscari of Venice, Cannaregio 873, 30121 Venice, Italy
e-mail: mcardin@unive.it

C. Perna, M. Sibillo (eds.), *Mathematical and Statistical Methods for Actuarial Sciences and Finance*, DOI 10.1007/978-3-319-05014-0_13,

## 2 Natural Risk Statistics

We assume that the behavior of a random loss $X$ is represented by a collection of data observation $\widetilde{x} = (x_1, x_2, \ldots, x_n) \in \mathbb{R}^n$ (could be empirical or subjective or both). A risk statistics $\widetilde{\rho}$ is a mapping from the data in $\mathbb{R}^n$ to a numerical value in $\mathbb{R}$. Next we postulate a set of axioms for the risk statistic $\hat{\rho}$.

**A1** Positive homogeneity and translation invariance:

$$\hat{\rho}(a\widetilde{x} + b\mathbf{1} = a\hat{\rho}(\widetilde{x}) + b \quad \text{if } \widetilde{x} \in \mathbb{R}^n,\ a \geqslant 0,\ b \in \mathbb{R},$$

where $\mathbf{1} = (1, 1, \ldots, 1) \in \mathbb{R}^n$.

**A2** Monotonicity:

$$\hat{\rho}(\widetilde{x}) \leqslant \hat{\rho}(\widetilde{y}), \quad \text{if } \widetilde{x} \leqslant \widetilde{y},$$

where $\widetilde{x} \leqslant \widetilde{y}$ if and only if $x_i \leqslant y_i, i = 1, \ldots, n$.

**A3** Comonotonic subadditivity:

$$\hat{\rho}(\widetilde{x} + \widetilde{y}) \leqslant \hat{\rho}(\widetilde{x}) + \hat{\rho}(\widetilde{y}), \quad \text{if } \widetilde{x} \text{ and } \widetilde{y} \text{ are comonotonic},$$

where $\widetilde{x}$ and $\widetilde{y}$ are comonotonic if and only if $(x_i - x_j)(y_i - y_j) \geqslant 0$, for any $i \neq j$.

**A4** Permutation invariance:

$$\hat{\rho}(x_1, x_2, \ldots, x_n) = \hat{\rho}(x_{i_1}, x_{i_2}, \ldots, x_{i_n}) \quad \text{for any permutation } (i_1, \ldots, i_n).$$

A risk statistic $\hat{\rho} : \mathbb{R}^n \to \mathbb{R}$ is called a natural risk statistic if it satisfies Axioms A1–A4. For a justification of the concept as well for a detailed study of the axioms we refer to [6].

## 3 OWA Risk Statistics

The problem of aggregating multiple numerical values into a single value is of considerable importance in many disciplines. The most commonly used aggregation is based on the weighted sum. In the ordered weighted averaging operators OWA developed by Yager in [10] the weights are assigned to the ordered values (i.e. to the smallest value, the second smallest and so on) rather than to the specific values. Since its introduction, the OWA aggregation has been successfully applied to many fields of decision making and recently considered in connection with risk measurement in [5]. Moreover the OWA operator allows us to model various aggregation functions from the maximum through the arithmetic mean to the minimum.

Now we consider OWA risk statistics that are risk statistics $\hat{\rho} : \mathbb{R}^n \to \mathbb{R}$ such that

$$OWA_{\widetilde{\omega}}(\widetilde{x}) = \sum_{i=1}^{n} \omega_i x_{(i)}$$

where for a vector $x \in \mathbb{R}^n$ we denote its elements ranked in ascending order as $x_{(1)} \leqslant x_{(2)} \cdots \leqslant x_{(n)}$ and $\widetilde{\omega} = (\omega_1, \omega_2, \ldots, \omega_n) \in [0, 1]^n$ with $\sum_{i=1}^{n} \omega_i = 1$.

It can be proved that a OWA risk statistic is a natural risk statistic that satisfy the following axiom:

**A5** Comonotonic additivity:

$$\hat{\rho}(\widetilde{x}+\widetilde{y}) = \hat{\rho}(\widetilde{x}) + \hat{\rho}(\widetilde{y}), \quad \text{if } \widetilde{x} \text{ and } \widetilde{y} \text{ are comonotonic.}$$

The following result characterize OWA risk statistics and prove that OWA risk statistics generate all risk statistics.

**Proposition 1** *A risk statistic is a OWA risk statistics if and only if satisfies axioms* A1, A2, A4 *and* A5.

*A risk statistic $\hat{\rho}$ is a natural risk statistic if and only if there exists a set of weights $W = \{\widetilde{\omega} = (\omega_1, \omega_2, \ldots, \omega_n)\} \subseteq \mathbb{R}^n$ such that*

$$\hat{\rho}(\widetilde{x}) = \sup_{\widetilde{\omega} \in W} OWA_{\widetilde{\omega}}(\widetilde{x}).$$

*Proof* A real operator defined on $\mathbb{R}^n$ is a Choquet integral with respect to a monotone measure if and only if it is monotone, positive homogeneous and comonotonic additive (see for example [9]) The operator is symmetric if and only the monotone measure is symmetric (see [7] for the definition of a symmetric measure and for the result). A Choquet integral with respect to a monotone measure is a OWA operator if and only if the measure is symmetric as is proved in [7]. The second part of the proposition follows from the main theorem in [6]. □

## 4 Generalized Gini Risk Statistics

We also consider the concept of majorization arising as a measure of diversity of the components of a $n$-dimensional vector. Majorization has been comprehensively treated by [3] and [8]. We aim to formalize the idea that the components of a vector $y$ are less "spread out" or "more nearly equal" than the components of $x$. The vector $x$ is said to majorize the vector $y$ which is denoted as $x \gtrsim y$, if

$$\sum_{i=1}^{k} x_{(i)} \leq \sum_{i=1}^{k} y_{(i)} \quad \text{for } k = 1, 2, \ldots, n-1 \text{ and } \sum_{i=1}^{n} x_i = \sum_{i=1}^{n} y_i. \tag{1}$$

Majorization is a partial ordering among vectors, which applies only to vectors having the same sum. It is a measure of the degree to which the vector elements differ. For example it can be easily shown that all vectors of sum $s$ majorize the uniform vector $u = (\frac{s}{n}, \ldots, \frac{s}{n})$. Intuitively, the uniform vector is the vector with minimal differences between elements, so all vectors majorize it.

A real function $f\colon \mathbb{R}^n \to \mathbb{R}$ is S-convex if $f(x) \geqslant f(y)$ when $x \gtrsim y$. $f$ is S-concave if $-f$ is S-convex. S-convex $i$ functions thus preserve majorization. We note also that a Schur increasing or decreasing function must be a symmetric function. Moreover a symmetric convex function is Schur increasing [8].

The following axiom consider majorization between data sets.

**A6** S-convexity:

$$\hat{\rho}(\widetilde{x}) \geqslant \hat{\rho}(\widetilde{y}), \quad \text{if } \widetilde{x} \gtrsim \widetilde{y}.$$

Then if we accept axiom A6 we consider more risky a situation in which data are more "spread out" . We can easily prove as in [2] the following result.

**Proposition 2** *A OWA risk statistics satisfies axioms A6 if and only if*

$$\omega_1 > \cdots > \omega_n.$$

It seems important to note that the OWA operators that satisfy the condition in Proposition 4 are associated with the classical Gini inequality measure (see [2] and the references therein) and are called generalized Gini functions. Finally, we intend to characterize natural risk statistics that satisfy axiom A6.

## References

1. Ahmed, S., Filipoviĉ, D., Svintland, G.: A note on natural risk statistics. Oper. Res. Lett. **36**, 662–664 (2008)
2. Aristondo, O., García-Lapresta, J.L., Lasso de la Vega, C., Marques Pereira, A.: Classical inequality index, welfare and illfare functions, and the dual decomposition. Fuzzy Sets Syst. **228**, 114–136 (2013)
3. Arnold, B.C.: Majorization and the Lorenz Order: A Brief Introduction. Springer, New York (1987)
4. Artzner, P., Delbaen, F., Eber, J.M., Heath, D.: Coherent measures of risk. Math. Finance **9**(3), 203–228 (1999)
5. Belles-Sampera, J., Merigó, J.M., Guillén, M., Santolino, M.: The connection between distortion risk measures and ordered weighted averaging operators. Insur. Math. Econ. **52**(2), 411–420 (2013)
6. Heyde, C.C., Kou, S.G., Peng, X.H.: What is a good external risk measure: bridging the gaps between robustness, subadditivity and insurance risk measures. Preprint (2007)
7. Marichal, J.L.: Aggregation of interacting criteria by means of the discrete Choquet integral. In: Calvo, T., Mayor, G., Mesiar, R. (eds.) Aggregation Operators: New Trends and Applications. Studies in Fuzziness and Soft Computing, vol. 97, pp. 224–244. Physica, Heidelberg (2002)
8. Marshall, A.W., Olkin, I.: Inequalities: Theory of Majorization and Its Applications. Academic Press, New York (1979)
9. Schmeidler, D.: Integral representation without additivity. Proc. Am. Math. Soc. **97**(2), 255–261 (1986)
10. Yager, R.R.: On ordered weighted averaging aggregation operators in multicriteria decision making. IEEE Trans. Syst. Man Cybern. **18**(1), 183–190 (1988)

# The Estimation of Standard Deviation of Premium Risk Under Solvency 2

**Rocco Roberto Cerchiara and Vittorio Magatti**

**Abstract** Solvency 2 Directive provides a range of methods to calculate the Solvency Capital Requirement (SCR). Focusing on the Standard Formula (SF) approach with Undertaking-Specific Parameters (USPs), the Technical Specifications (TS) of Quantitative Impact Study 5 (QIS5) describes a subset of the SF market parameters (standard deviations) that may be replaced by USPs, in order to calculate the SCR deriving from Premium Risk, using three different standardised methods. Compared to the existing literature and practice, this paper innovates in that this standard deviation will be calculated using a Partial Internal Risk Model (PIRM), based on Generalised Linear or Additive Models (GLM or GAM), showing how the techniques usually developed for premium calculation could be useful for this goal.

**Keywords** Solvency 2 · Premium risk · Undertaking specific parameters

## 1 Introduction

GLM and GAM can be used in order to define an estimated Aggregate Claim Amount (ACA) as a function of the tariff variables detectable in the insurance contract and consequently a propensity of each insured to produce a loss for the undertaking. Pricing Staff is interested on determining the expected value of the ACA (risk neutral) and Risk Management could identify the volatility (real world) in the pricing process calculating the moments and/or the percentiles of the estimated distribution of the ACA with the same model. Perimeter of the approach we propose in this paper will cover personal line insurance or generally product priced using regression techniques. If GLM represents a benchmark within this technical framework, GAM is an interesting alternative for its non-parametric or semi-parametric

R.R. Cerchiara (✉)
University of Calabria, Rende Cosenza, Italy
e-mail: cerchiara@unical.it

V. Magatti
University of Rome La Sapienza, Rome, Italy
e-mail: vittorio.magatti@uniroma1.it

C. Perna, M. Sibillo (eds.), *Mathematical and Statistical Methods for Actuarial Sciences and Finance*, DOI 10.1007/978-3-319-05014-0_14,

structure and furthermore when the distribution of the aggregate claim amount for the tariff variables are not linear. This paper will show a summary of the outputs, statistics and graphical analysis of residuals necessary to validate the optimum GLM and GAM, but also to exhibit which model predicts better the expected value of the ACA. Using models outlined above, a comparison between the SF market parameters, USPs (see Sect. 2 for more details) and the standard deviation of the PIRM will be shown in Sect. 3, regarding a case study for the Line of Business Motor Third Party Liability (MTPL). More details will be given in the extended version of this short paper.

## 2 Undertaking Specific Parameters

CEIOPS (now EIOPA) in the 5th Quantitative Impact Studies (QIS5, see [2]) showed the option of using USPs for Premium Risk and Reserving Risk (see [3] for a market analysis of USPs for Italian Market). Three alternative methods are presented in QIS5 about the calculation of USPs for Premium Risk. The first two methods are similar as approach and in the assumptions. An insurance company could use those methods when the loss is proportional to premium and if Company has a different but constant Expected Loss Ratio (ELR). Furthermore, Methods 2 is based on the LogNormal distribution of the losses. The third method is a frequency/severity approach. Whereas the first two methods are influenced by the volatility of Earned Premium and/or of the ELR, Method 3 is influenced by the exposure and/or in the number of claim and it is more onerous than the other one concerning the input. USPs should be calibrated on the basis of internal data and the use of them requires supervisory approval. Data in particular used for the calculation of USPs should be complete, accurate and appropriate (see [2]).

## 3 Partial Internal Model Specification

As reported in [5], Member States shall ensure that insurance or reinsurance undertakings may calculate the Solvency Capital Requirement using a Full or Partial Internal Model as approved by the Supervisory Authorities. Since the Premium Risk is defined as the risk due to error in the assumptions, models or methods used to solve a pricing problems, the solution we propose in this paper is to stress the opportunity that Risk Management has to find this information directly from the model used in the Pricing Staff to determine the premium. Generally this approach is valid for personal line guarantees and/or for product priced using actuarial methods. In order to define and solve a pricing problem, Pricing Staff will have to study the random variable (r.v.) Aggregate Claim Amount (ACA):

$$ACA = \sum_{i=1}^{N} Y_i . \tag{1}$$

Where $N$ is the r.v. number of claims and $Y$ the r.v. amount of claim. In [4, 6] and [8], it is demonstrated that, if $Y_i$ for $i$ in $(1,\ldots,N)$ are independent and identically distributed, the premium $P$ is equal to:

$$E(ACA) = E(N) \cdot E(Y) = P. \tag{2}$$

Observing the historical number and amount of the incurred claims summarized for each risk profile (i.e. insured), Risk Management should set a real world global level of those figures, according with the Actuarial Function. The biggest difference with respect to the calculation of the premium concerns the time horizon. Actually Risk Management have to determine a SCR in one-year horizon and not at ultimate cost. For this reason, we update the observed incurred amount of claim to reach the ultimate one-year view at policy level ($ACA_{1-yr}$). Article 122 of the Solvency Directive (see [5]) allows the (re-) insurance undertaking to use a different time period or risk measure to calculate the Solvency Capital Requirement with a PIRM, in a manner that provides policy holders and beneficiaries with a level of protection equivalent. Solution we present in this paper is to calculate:

$$\sigma_{(prem,lob)}^{PIRM} = \sigma(ACA_{1-yr}). \tag{3}$$

Two pricing models are used to identify the moments of estimated $ACA_{1-yr}$ for each insured: Generalized Linear Model (GLM) and Generalized Additive Model (see [7, 9] and [10] for more details). Process we follow to model the $ACA_{1-yr}$ with GAM consists of two parts: a non-parametric fitting for all the risk factors to understand what are the regressors distributed not linearly against to $ACA_{1-yr}$ and a semi-parametric step in which the regressors identified before are fitted without an assumption about the error structure, whilst all other rating factors are fitted with the same assumption of GLM: Poisson distribution for the r.v. $N$ number of claims (i.e. frequency) and Gamma distribution for the r.v. $Y$ the r.v. amount of claim (i.e. severity). Within this probabilistic framework, assuming independence between r.v. $N$ and $Y_i$, the convolution between the frequency and severity models is distributed as a Gamma under a Poisson compound process (see [4] and [8]). Extending the previous formula, the solution we present in this paper becomes:

$$\sigma_{(prem,LoB)}^{PIRM} = \sigma(ACA_{1-yr}) = \sqrt{\frac{E(ACA_{1-yr})^2}{\phi}} \tag{4}$$

where $\phi$ is the scale parameter (GLM and GAM calculate this parameter when fit the $ACA_{1-yr}$). To appreciate the differences between market-wide parameters, USPs and PIRM, we develop a case study for an hypothetical MTPL portfolio for cars: a large number of constraints have been considered in out IT procedure in order to define risk profiles with simulated numbers and amounts of claims coherent with the statistics of the Italian Market (see [1]). According this procedure, we define a portfolio of a medium size Italian company, assuming it was authorized from Supervisory at least 15 years ago and we set also the global level in one-year horizon of amount of claims for each risk profile underwritten between 2009–2011. Fitting a non-parametric GAM, only the policy duration is distributed not linearly against to $ACA_{1-yr}$, both in frequency ($N$) and severity ($Y$) models. Furthermore, after the

**Table 1** Summary of results

| Market-Wide | USP-Met. 1 | USP-Met. 2 | USP-Met. 3 | GLM | GAM |
|---|---|---|---|---|---|
| 10.0 % | 8.0 % | 7.8 % | 9.1 % | 4.0 % | 3.9 % |

convolution between frequency and severity model, assuming Gamma distribution, GAM it seems to be the candidate PIRM. However, GLM produces results very close to GAM and, for this reason, we will present below the volatility of the Premium Risk for both models with a comparison with the three QIS5 methods (see Table 1).

Due to the correlation between Premium and Reserving risk (see [2]), we cannot affirm that GLM or GAM can reduce the SCR for the Underwriting Risk for every company, but PIRMs allow a considerable saving for this (hypothetical) Insurance Company considering only Premium Risk. PIRMs have to be submitted for supervisor pre-approval and so are more onerous in terms of resources and costs respect to Standard Formula or USPs, but in this paper we would remark the opportunity under a PIRM that a company could create a joint approach between Pricing Staff, Actuarial Function and Risk Management in order to define and monitor the underwriting risk and/or which are exposed to, strictly linked to SCR definition.

## References

1. ANIA: Italian Insurance 2010–2011 (2009). http://www.ania.it/it/index.html
2. CEIOPS: QIS5 Technical Specifications, Brussels (2010). http://ec.europa.eu
3. Cerchiara, R.R., Santoni, A.: Analisi di Mercato: Undertaking Specific Parameters. ANIA, Rome (2010). http://www.ania.it/it/index.html
4. Daykin, C.D., Pentikäinen, T., Pesonen, M.: Practical Risk Theory for Actuaries. Chapman & Hall, London (1994)
5. European Parliament: legislative resolution of 22 April 2009 on the amended proposal for a directive of the European Parliament and of the Council on the taking-up and pursuit of the business of Insurance and Reinsurance (2009)
6. Gigante, P., Picech, L., Sigalotti, L.: La Tariffazione nei Rami Danni con Modelli Lineari Generalizzati. EUT editore, Trieste (2010)
7. Hastie, T.J., Tibshirani, R.J.: Generalized Additive Models. Chapman & Hall, New York (1990)
8. Klugman, S., Panjer, H., Willmot, G.: Loss Models—From Data to Decisions, 3rd edn. Wiley, New York (2008)
9. Nelder, J.A., Wedderburn, R.W.M.: Generalized linear models. J. R. Stat. Soc. A **135**, 370–384 (1972)
10. Ohlsson, E., Johansson, B.: Non-life Insurance Pricing with Generalized Linear Models. EAA Series Textbook. Springer, Berlin (2010)

# The Solvency Capital Requirement Management for an Insurance Company

**Mariarosaria Coppola and Valeria D'Amato**

**Abstract** Longevity risk plays a central role in the insurance company management since only careful assumptions about future evolution of mortality phenomenon allows the company to correctly front its future obligations. According to Solvency II longevity risk represents a sub-module of the underwriting risk module in the regulatory standard formula. In this paper we examine the adequacy of the shock's structure suggested by the standard formula studying its impact on the solvency capital requirements and liabilities at different ages. In particular, we propose an alternative to the regulatory standard model represented by a flexible internal model. The innovative approach hinges on a stochastic volatility model and a so-called coherent risk measure as the expected shortfall. An empirical analysis is provided.

**Keywords** Solvency capital requirement · Longevity risk · Longevity shocks · Expected shortfall

## 1 Introduction

The dynamic management process of an insurance company business is affected by many different kinds of risks. In particular, in the case of life insurance we can recognize two main risks: the investment risk and the demographic risk. The former derives by the random fluctuations of the financial market. The latter can be split into insurance risk, due the random deviation of the number of deaths from its expected value, and longevity risk deriving from the improvement in mortality rates. Longevity risk plays a central role in the insurance company management since only careful assumptions about future evolution of mortality phenomenon allows the company to correctly front its future obligations. Longevity risk represents

M. Coppola (✉)
Department of Political Sciences, Federico II University, via L. Rodinò 22, 80138 Naples, Italy
e-mail: m.coppola@unina.it

V. D'Amato
Department of Statistics and Economics, Campus di Fisciano, University of Salerno, via Giovanni Paolo II 132, 84084 Fisciano (SA), Italy
e-mail: vdamato@unisa.it

C. Perna, M. Sibillo (eds.), *Mathematical and Statistical Methods for Actuarial Sciences and Finance*, DOI 10.1007/978-3-319-05014-0_15,

a sub-module of the underwriting risk module in the Solvency II standard formula. According to this formula Solvency Capital Requirement for Longevity Risk (SCR-Long) should be calculated as the change in the net asset value (NAV) due to a pre-specified longevity shock. This shock is actually represented by a 20 % permanent reduction of mortality rates for each age and contract linked to longevity risk. To correctly calculate the solvency capital requirements we evaluate at the beginning of each year the amount of capital that the insurer need to meet its future obligations year by year till the contract will be in force. We examine the adequacy of the shocks structure suggested by the standard formula studying its impact on the SCRLong and liabilities at different ages. A constant reduction of mortality rates for all ages could cause a result of underestimation or overestimation of longevity risk changing with the different considered ages. We discuss about a framework based on an age-dependent longevity shock. It seems to be the only way allowing the insurer to correctly calculate the solvency capital requirement to front its future obligations. We propose an alternative to the regulatory standard model represented by a flexible internal model. To avoid biased allocation of capital the mathematical formulation of the problem considers that the liabilities of the life insurance company are related to the longevity phenomenon volatility and the volatility evolution is described by an appropriate stochastic process. The innovative approach hinges on the stochastic volatility model and a so-called coherent risk measure as the expected shortfall.

The layout of the paper is organised as follows. In Sect. 2 a stochastic model for representing the volatility of the longevity shocks is proposed. Section 3 is devoted to some remarks on the innovative volatility-adjusted internal model.

## 2 The Longevity Shock Model

One of the main management problems that actuaries have to front is the determination of the Solvency Capital Requirements, which represents the main contribution in the Solvency II framework. The SCR is set up to prevent policyholders against unforeseen losses [2]. According to this regulation, the SCR can be computed by a standard formula or an internal model. The basic principle is that the SCR is determined as the 99.5 % Value at risk (VaR) of the Available Capital over one-year time horizon [1]. The SCR for longevity risk under Solvency II standard formula is defined as the net change in Net Asset Value (NAV) due to a permanent longevity shock equal to 20 % of the mortality rates for each age under a specific survival scenario at time $t = 0$. The adequacy of the one-off shock structure that is equal for all ages and maturities appears to be less appropriate than a gradual change in mortality rates. In particular, some authors like [6] come to conclusion that the shock scenario referred to by the standard formula can be far away from the actual experience of the insurer, and thus may lead to a biased allocation of capital. In light of the previous considerations, we propose to model the longevity shock as a function of mortality rate volatility. To this aim, we recognize two kinds of mortality rate volatility: the time volatility $v_{x_j}(t_j)$ and the age volatility $v_{t_j}(x_j)$. In the former case, mortality rate volatility expresses the variability of mortality as the time varies. In the latter one, it represents the variability of mortality as the age varies. Furthermore, in a dynamic approach of the longevity phenomenon, the shock also depends on the calendar year. On the basis of these issues, we suppose that the mortality rate volatility evolution is described by the stochastic process $V_t$ and the longevity shock $LS$ is given by:

$$LS = f(V_t, t) \tag{1}$$

resulting as a function of the stochastic mortality rate volatility $V_t$ and the time $t$. Let us suppose the PDE for longevity shock is:

$$\frac{\partial LS(v,t)}{\partial t} + a(v,t_j)\frac{\partial LS(v,t)}{\partial v} + \frac{1}{2}b^2(v,t) \times \frac{\partial^2 LS(v,t)}{\partial v^2} = 0 \tag{2}$$

being $LS$ a solution of the deterministic PDE, under the terminal condition: $LS(v,T) = h(v)\ \forall v \in R$, where $a(v,t)$, $b^2(v,t) : [0,T] \times R \to R$, $h(v) : R \to R$ are given functions such that $\{V_t\}_{t\geq 0}$ is the solution to the SDE:

$$dV_s = a(V_s,s)ds + b(V_s,s)dz(t) \quad \forall s \in [t,T] \tag{3}$$

being $dz(t)$ a Wiener process and $V_t = v$.

According to Feynman-Kac theorem if: $E[\int_t^T |b(V_s,s)\frac{\partial}{\partial v}LS(V_s,s)|^2 ds] < \infty$ the solution to PDE has the following form:

$$LS(t,v) = E\big[h(V_t)|V_t = v\big] \quad \forall (t,v) \in [0,T] \times R. \tag{4}$$

Formula (4) represents the stochastic evolution in time of the mortality rate volatility in the general case. It can be characterized in specific different way. For example we can consider that it is governed by an Ornstein-Uhlenbeck process as in [8] and [7], so we can rewrite (4) as: $dV_s = -\delta(V_s - \Theta)ds + kdz(t)$. Alternatively we can suppose that the mortality rate volatility is described by the square-root process as in [4]: $dV_t = k[\vartheta - V_t]dt + \sigma\sqrt{V_t}dz(t)$.

The Expected Shortfall ($ES$) allows to determine the expected loss incurred in the $\alpha = A\ \%$ [3] worst cases of our portfolio. It is given by:

$$ES_\alpha = \frac{1}{\alpha}\int_0^\alpha VaR_{1-\gamma}(X)\,d\gamma. \tag{5}$$

In our case $X$ represents the loss variable defined as follows:

$$X = \left(\frac{BEL_t - CF_t}{1 + i(t-1,t)}\right) - BEL_{t-1} \tag{6}$$

where $CF_t$ denotes the company's stochastic cash flow at time $t$, and $BEL_t$ is the *best estimate* of liabilities at time $t$.

## 3 Remarks

Insurance companies allocate capital through their yearly budgeting process. It depends on international guidelines and internal decisions about different lines of business. To achieve an effective, dynamic and forward-looking allocation of capital a consistent approach is required. A consistent approach does not often correspond to the standard model proposed in Solvency II. Under Solvency II, the capital requirements are evaluated for the separate risk classes. In particular, the capital charge for longevity risk results by the net change in Net Asset Value due to a constant longevity shock equal to 20 %. The empirical evidences show an age-dependent shock which highlights the structural shortcoming of the standard model, suggesting another more appropriate configuration for calculating the solvency capital requirements. The alternative to the standard model is represented by an internal model. According to Groupe Consultatif CEA Glossary, it consists in a risk measurement system developed by an insurer to analyse the overall risk position, to quantify the risks and to determine the economic capital required to meet those risks. To make effective an insurer's internal model in risk and capital management it has to be fully embedded into the risk strategy and operational processes of the insurer [5]. In this research we propose an internal volatility-adjusted model by representing the longevity shock on the basis of a stochastic volatility model and a coherent risk measure as the expected shortfall. For the company's own flexibility, the SCRs are calculated by using various risk measures, confidence levels and time horizons, so that other risk indexes not recognized in the regulatory model are properly included.

## References

1. Borger, M.: Deterministic shock vs. stochastic value-at-risk: an analysis of the solvency II standard model approach to longevity risk. Blätter DGVFM **31**, 225–259 (2010)
2. Bouma, S.: Risk Management in the Insurance Industry and Solvency II. Capgemini Compliance and Risk Management Centre of Excellence (2006)
3. CEIOPS, QIS 5, Technical Specifications (2010). https://eiopa.europa.eu/consultations/qis/quantitative-impact-study-5/technical-specifications/index.html
4. Cox, J.C., Ingersoll, J.E., Ross, S.: A theory of the term structure of interest rates. Econometrica **53**(2), 385–407 (1985)
5. IAIS, International Association of Insurance Supervisors: global reinsurance. Market Report (2008)
6. Olivieri, A., Pitacco, E.: Solvency requirements for life annuity: some comparisons. G. Ist. Ital. Attuari **LXXI**(1–2), 59–82 (2008)
7. Schobel, R., Zhu, J.W.: Stochastic volatility with an Ornstein-Uhlenbeck process: an extension. Eur. Finance Rev. **3**, 23–46 (1998)
8. Stein, J., Stein, E., Price, S.: Distributions with stochastic volatility: an analytic approach. Rev. Financ. Stud. **4**, 727–752 (1991)

# Direct Multi-Step Estimation and Time Series Classification

**Marcella Corduas**

**Abstract** The AR metric represents a consolidated model-based approach for time series classification. The goodness of the final classification may of course be affected by the misspecification of the models describing the observed time series. This article investigates whether a direct multi-step estimation approach can shed some more light on time series comparison.

**Keywords** AR metric · Time series classification · Adaptive estimation · Direct multi-step estimation

## 1 Introduction

Direct multi-step estimation (DMS) has attracted considerable attention in recent years as a strategy for improving multi-step ahead forecasts (see [13] and [4] for an extensive review). In particular, among the earlier contributions, Tiao and Xu [16] extended the results by Cox [8] and showed that direct multi-step estimation can lead to more efficient forecasts when the model is misspecified. In the same line, Tiao and Tsay [15] discussed the use of multi-step estimation for forecasting ARFIMA processes. According to those authors, the rationale for the use of multi-step (adaptive) forecasting is that all statistical models are imperfect representations of reality and that local approximations are more relevant than global ones when the objective of the analysis is forecasting. For this reason, different models should be fitted for each forecast. The debate is still alive and various controversial works have been published about the performance of iterated and adaptive forecasting [9].

The focus of the present article is on model based classification for time series. A consolidated approach, proposed by Piccolo ([10, 11]), is based on the comparison of the data generating processes by means of the AR metric. Under suitable assumptions, the AR metric can be interpreted as a distance measure between the forecasting functions associated to each generating process. The goodness of the final classification may of course be affected by the misspecification of the ARIMA

M. Corduas (✉)
Department of Political Sciences, University of Naples Federico II, Naples, Italy
e-mail: corduas@unina.it

C. Perna, M. Sibillo (eds.), *Mathematical and Statistical Methods for Actuarial Sciences and Finance*, DOI 10.1007/978-3-319-05014-0_16,

models describing the observed time series. Thus, it is worth investigating whether the DMS approach can improve time series comparison. The article is organized as follows: firstly, the methodology proposed in the present paper is outlined by introducing the AR metric and its extension. Secondly, an empirical case study and some remarks about further methodological developments are discussed.

## 2 An Extension of the AR Metric for ARIMA Models

The class of Gaussian *ARIMA* processes provides a useful parsimonious representation (see [3]) for linear time series. Specifically, $Z_t \sim ARIMA(p,d,q)$ is defined by:

$$\phi(B)\nabla^d Z_t = \theta(B)a_t, \tag{1}$$

where $a_t$ is a Gaussian White Noise (WN) process with constant variance $\sigma^2$, $B$ is the backshift operator such that $B^k Z_t = Z_{t-k}$, $\forall k = 0, \pm 1, \ldots$, the polynomials $\phi(B) = 1 - \phi_1 B - \cdots - \phi_p B^p$ and $\theta(B) = 1 - \theta_1 B - \cdots - \theta_q B^q$, have no common factors, and all the roots of $\phi(B)\theta(B) = 0$ lie outside the unit circle. Moreover, we assume that the time series has been preliminary transformed in order to improve Gaussianity, to deal with non-linearities, to reduce asymmetry, and to remove any outlier or deterministic components.

A distance criterion which compares the forecasting functions of two *ARIMA* models given a set of initial values was proposed by Piccolo ([10, 11]). In particular, assuming that $Z_t$ is a zero mean invertible process which admits the $AR(\infty)$ representations: $\pi(B)Z_t = a_t$, the $\pi$-weights sequence and the WN variance completely characterize $Z_t$ (given the initial values). Hence, a measure of structural diversity between two *ARIMA* processes with given orders, $X_t$ and $Y_t$, can be defined as:

$$D_{AR} = \sqrt{\sum_{j=1}^{\infty} (\pi_{xj} - \pi_{yj})^2}. \tag{2}$$

The WN variances are not included in the distance formulation since they depend on the units of measurement. The criterion has been widely experimented (see [12] for a review) and the asymptotic properties have been derived under general assumptions ([5, 7]).

Although the AR metric is defined for ARIMA processes in the following, for the sake of simplicity, we restrict our attention to stationary processes. Moreover, we do not consider the problem of order selection (see, for instance, [1, 2, 14]) but we assume that the structure of the time series model does not change with the forecast horizon.

Most applications of the AR metric involve a large number of time series to be compared. In such a case, each time series is treated as a realization of an $AR(\infty)$ process which is approximated by an $AR(k)$ model. This approximate representation of the unknown generating process becomes the basis for the evaluation of the AR distance. The selection of the order $k$ is performed by means of an automatic criteria such as BIC. In this respect, given the initial values, we can consider two approaches for producing a forecasting function.

The first one relies on the standard least squares approach which estimates the unknown AR coefficients by minimizing the sum of squares of one-step ahead prediction errors: $\sum_{t=k}^{n-1}(Z_{t+1} - \phi_1 Z_t - \cdots - \phi_k Z_{t-k+1})^2$, where $n$ is the number of observations. This produces a set of estimated coefficients which are used to construct the $h$-step forecasts (for $h = 1, 2, \ldots, m$) by repeatedly iterating the autoregression and by using the plug-in principle to replace unknown values.

In the second alternative, instead, the $h$-step forecasts are constructed by fitting the autoregression separately for each lead time. Specifically, the AR coefficients are obtained by minimizing: $\sum_{t=k}^{n-h}(Z_{t+h} - \phi_1^h Z_t - \cdots - \phi_k^h Z_{t-k+1})^2$ for $h = 2, 3, \ldots, m$.

We propose to use the set of estimated coefficients: $(\phi_1^h, \ldots, \phi_k^h)$, $h = 1, \ldots, m$ in order to characterize a time series at each lead time $h$. Thus, the dissimilarity measure between $X_t$ and $Y_t$ will be defined as:

$$\widetilde{D}^2_{(h)} = \sum_{j=1}^{k} \left(\phi_{xj}^h - \phi_{yj}^h\right)^2 \quad h = 1, \ldots, m. \tag{3}$$

Finally, the overall dissimilarity measure can be introduced as:

$$\widetilde{D}^2_{A.(h)} = \frac{1}{m} \sum_{h=1}^{m} D^2_{(h)}. \tag{4}$$

The dissimilarity will be zero if and only if, given a set of initial values, the AR models characterizing the two time series will produce the same *adaptive* multi-step ahead forecasts.

## 3 An Empirical Application

We applied the proposed technique in order to compare the dynamics of 80 time series consisting of 2000 observations of mean daily discharge of streamflow rivers. This dataset was deeply discussed by Corduas [6] and represents a useful benchmark since the dynamics of this type of series may be affected by long memory components and thus the use of AR models may lead to misspecified structures. This is a common situation which may arise in other fields such as the analysis of financial data. The proposed dissimilarity measure seems to improve the representation provided by the complete linkage clustering method with respect to the results produced by using $D^2_{AR}$. Specifically, the representation shows two large clusters and remain unchanged for most time series. However, the new dissimilarity measure enhances: (i) a larger number of isolated elements; (ii) a number of time series which move from one cluster to the other due to an improved description of the inertial components that typically characterize streamflow time series.

The increased selectiveness of the proposed approach is very promising and pave the way to further studies. Specifically, the statistical properties of the new distance criterion have to be investigated in order to make it an effective tool. The main advantage of the AR metric is, in fact, related to its statistical properties which allows to set the time series comparison in an inferential framework.

**Acknowledgements** This work was financially supported by PRIN project 2013-2015: "La previsione economica e finanziaria: il ruolo dell'informazione e la capacità di modellare il cambiamento".

## References

1. Bhansali, R.: Asymptotically efficient autoregressive model selection for multistep prediction. Ann. Inst. Stat. Math. **48**, 577–602 (1996)
2. Bhansali, R.J.: Parameter estimation and model selection for multistep prediction of a time series: a review. In: Gosh, S. (ed.) Asymptotics, Nonparametrics, and Time Series, pp. 201–258. CRC Press, Boca Raton (1999)
3. Brockwell, A.P.J., Davies, R.A.: Time Series: Theory and Methods, 2nd edn. Springer, New York (1991)
4. Chevillon, G.: Direct multi-step estimation and forecasting. J. Econ. Surv. **21**, 746–785 (2007)
5. Corduas, M.: La metrica autoregressiva tra modelli ARIMA: una procedura in linguaggio GAUSS. Quad. Stat. **2**, 1–37 (2000)
6. Corduas, M.: Clustering streamflow time series for regional classification. J. Hydrol. **407**, 73–80 (2011)
7. Corduas, M., Piccolo, D.: Time series clustering and classification by the autoregressive metric. Comput. Stat. Data Anal. **52**, 1860–1872 (2008)
8. Cox, D.R.: Prediction by exponentially weighted moving averages and related methods. J. R. Stat. Soc. B **23**, 414–422 (1961)
9. Marcellino, M., Stock, J.H., Watson, M.: A comparison of direct and iterated multistep AR methods for forecasting microeconomic time series. J. Econom. **135**, 499–526 (2006)
10. Piccolo, D.: Una topologia per la classe dei processi ARIMA. Statistica **XLIV**, 47–59 (1984)
11. Piccolo, D.: A distance measure for classifying ARIMA models. J. Time Ser. Anal. **11**, 153–164 (1990)
12. Piccolo, D.: The autoregressive metric for comparing time series models. Statistica **LXX**, 459–480 (2010)
13. Proietti, T.: Direct and iterated multistep AR methods for difference stationary processes. Int. J. Forecast. **27**, 266–280 (2011)
14. Shibata, R.: Asymptotically efficient selection of the order of the model for estimating parameters of a linear process. Ann. Stat. **8**, 147–164 (1980)
15. Tiao, G.C., Tsay, R.S.: Some advances in non-linear and adaptive modelling in time-series analysis. J. Forecast. **13**, 109–131 (1994)
16. Tiao, G.C., Xu, D.: Robustness of maximum likelihood estimates for multi-step predictions: the exponential smoothing case. Biometrika **80**, 623–641 (1993)

# Alternative Assessments of the Longevity Trends

**Valeria D'Amato, Steven Haberman, Gabriella Piscopo, and Maria Russolillo**

**Abstract** The improvement of the longevity trend constitutes a great challenge for society. The long-term social and economic impact on health and care services as well as on the provision of pensions, annuities and insurance requires to accurately understand the uncertainty in the future evolution of life expectancy. The most popular and widely used model for projecting longevity is the well-known Lee Carter model. This study considers recent model enhancements in the present setting by comparing their main benefits and drawbacks.

**Keywords** Lee Carter model · Variance reduction techniques · Sieve bootstrap · Vector auto-regression

## 1 Background

Over the last 150 years there has been an improvement in mortality for adults and a decrease most significantly impacting the elderly. This evolution is observable in the majority of industrialised countries, y looking at the data available at the Human Mortality Database (HMD [7]). In Italy, males have realised an average annual gain of 3 months or a quarter of life expectancy at birth since 1960; and 4.2 months on the

V. D'Amato (✉) · M. Russolillo
Department of Statistics and Economics, Campus di Fisciano, University of Salerno, Fisciano (SA), Italy
e-mail: vdamato@unisa.it

M. Russolillo
e-mail: mrussolillo@unisa.it

S. Haberman
Faculty of Actuarial Science and Insurance, Cass Business School, City University, London, UK
e-mail: s.haberman@city.ac.uk

G. Piscopo
Department of Economics, University of Genoa, Genoa, Italy
e-mail: piscopo@economia.unige.it

C. Perna, M. Sibillo (eds.), *Mathematical and Statistical Methods for Actuarial Sciences and Finance*, DOI 10.1007/978-3-319-05014-0_17,

last ten years [2]. Other trends are also emerging, such as the cohort effect, which is especially evident in the United Kingdom where the generation born between 1925 and 1945 has experienced impressive improvements in longevity, with improvement rates well above those born in previous generations, and even those born afterward. This cohort effect is not homogeneous across countries or periods. The most popular approaches in the actuarial field is the Lee-Carter model ([8]; hereafter LC). This model has been extended to capture the main features of the dynamics of mortality rates as in [3] and in [9–12].

Nevertheless, it is important to quantify the uncertainty in projections through the computation of prediction intervals. In order to improve the predictions of survival probabilities we propose different stochastic approaches represented in the LC framework by remarking the main advantages and critical issues of each of them. In particular we develop integrated Variance Reduction Techniques (VRTs), discussing the main features of the advances. In Sect. 3 we stress the main advantages in comparison with a dependence-based approach in the Lee Carter setting (as proposed in [6]).

## 2 Methodological Approaches

Among the methods that are commonly used for improving a simulation process on the speed and efficiency, i.e. the VRT's, the stratified sampling is a multiple steps approach where basic idea behind is to take advantage from the a-priori information. The a-priori information is used to select the stratification variable for performing a significant grouping of the elements of the available dataset into a disjoint subsets of the whole space, the so-called strata, from which the sample was drawn. The determination how many strata are suitable corresponds to the question whether an advantage from stratification with the chosen variable can be expected at all.

Let consider the case of the a rectangular mortality data array $(d_{xt}, e_{xt})$, comprising the numbers of deaths, modeled as independent Poisson responses $d_{xt}$, being $e_{xt}$ the exposures to the risk of death, in combination with the log-bilinear structure for the force of mortality

$$m_{xt} = \exp(\alpha_x + \beta_x \kappa_t) \tag{1}$$

where respectively $\alpha_x$ represents the general shape of the age-specific mortality profile and $\beta_x$ the age-specific component indicating which rates decline rapidly and which slowly over time in response to changes in time-varying parameter $\kappa_t$.

In the present setting, the number of groups (strata) to be considered has to be determined by the characteristics of the array under consideration. In particular strata per single ages can be considered. The idea behind is to integrate basic VRTs in order to take advantage of their respective complementary properties, leading to a variance reduction of the estimates wider than using the single constituent technique. Thus, the integrated VRT's can achieve high efficiency gains even though the respective constituent basic VRT's are not applied under optimal conditions. Relatively little work has been directed towards integrating some of VRT's in an overall scheme that can exploit various sources of efficiency improvements simultaneously [1], while they have not been used so far in the context of LC projections. In particular, we produce stratified bootstrap frames using the stratified sampling techniques. Then, for each stratified bootstrap frame the antithetic frame is generated. The simulation scheme works as follows. First, the stratification is performed. Then the semi-parametric bootstrap is implemented, for each stratum in the so-called original samples and simultaneously in the antithetic one. We observe that the antithetic estimator is obtained by combining the two estimators: the original and the antithetic as well. In the event of negative correlation between the estimators, arising from the samples being dependent, a variance reduction is realized. To exploit various sources of efficiency improvement simultaneously, we arrange the following scheme:

1. Stratification process of the number of deaths at age $x$ at time $t$, $d_{xt}$, into relatively homogeneous subgroups according to the Stratified Sampling Technique (SS);
2. Bootstrapping $m = 1, 2, \ldots, M$ samples from each of the strata simulating responses $d^*_{xt}$ drawn from $Poi(\hat{d}_{xt})$. We will refer to $m = 1, 2, \ldots, M$ as Original Samples;
3. Generation of the respective highly negative correlated $m = 1, 2, \ldots, M$ Antithetic Samples for each of the strata, according to the Antithetic Variables Technique (AV);
4. Fitting the log-bilinear structure to $d^*_{xt}$ for obtaining the model parameter vector $\theta^*_o = (\hat{\alpha}^*_x, \hat{\beta}^*_x, \hat{\kappa}^*_t)$ for $m = 1, 2, \ldots, M$ original samples and simultaneously for the respective antithetic samples $\theta^*_A = (\hat{\alpha}^*_x, \hat{\beta}^*_x, \hat{\kappa}^*_t)$;
5. Projecting mortality estimates on the basis of the $\theta^* = (\hat{\alpha}^*_x, \hat{\beta}^*_x, \hat{\kappa}^*_t)$ derived from the application of the antithetic reduction properties.

Synthetically, the simulation scheme works as follows. First, the stratification is performed. Then the semi-parametric bootstrap is implemented, for each stratum in the so-called original samples and simultaneously in the antithetic one. We observe that the antithetic estimator is obtained by combining the two estimators: the original and the antithetic as well. In the event of negative correlation between the estimators, arising from the samples being dependent, a variance reduction is realized.

Nevertheless, the interactions between age and time is neglected in the proposed algorithm. As shown in [5, 6] when the dependency risk is ignored, an inefficient risk management is obtained.

The predictor structure proposed in [6] takes up the idea of first fitting Lee Carter parametric model, and then re-sampling the centred residuals, according to an autoregressive approximation for generating bootstrap replications of the data.

## 3 Remarks

Following [4], a significant key criterion for detecting whether any stochastic mortality model is "suitable" or not "forecast levels of uncertainty and central trajectories should be plausible and consistent with historical trends and variability in mortality data".

The aforementioned algorithms represent enhancements in LC respect. They produce more accurate confidence intervals (CI's) for longevity projections. The VRTs approach leads to narrower CI's, even if a smaller number of sources of risk is taken into account. The risk management from insurance companies point of view could be less effective.

On the contrary when the overall sources of risk are taken into account, including the dependency risk, the uncertainty measurement is more a reliable. Nevertheless the CI's will be wider.

More details about the impact of the approaches under consideration on the actuarial evaluations will be given in the extended version of this short paper.

## References

1. Avramidis, A.N., Wilson, J.R.: Integrated variance reduction strategies for simulation. Oper. Res. **44**(2), 327–346 (1996)
2. Axa Papers, No.1 Longevity, June (2010)
3. Brouhns, N., Denuit, M., van Keilegom, I.: Bootstrapping the Poisson log-bilinear model for mortality forecasting. Scand. Actuar. J. **3**, 212–224 (2005)
4. Cairns, A.J.G., Blake, D., Dowd, K., Coughlan, G.D., Epstein, D., Khala Allah, K.: Mortality density forecasts: an analysis of six stochastic mortality models. The Pensions Institute, Discussion Paper 0801. Cass Business School (2008)
5. D'Amato, V., Haberman, S., Piscopo, G., Russolillo, M.: Modelling dependent data for longevity projections. Insur. Math. Econ. **51**, 694–701 (2012). doi:10.1016/j.insmatheco
6. D'Amato, V., Haberman, S., Piscopo, G., Russolillo, M.: Computational framework for longevity risk management. Comput. Manag. Sci. (2013). doi:10.1007/s10287-013-0178-2. Print ISSN 1619-697X, Online ISSN 1619-6988
7. Human Mortality Database. University of California, Berkeley (USA), and Max Planck Institute for Demographic Research (Germany). Available at www.mortality.org or www.humanmortality.de
8. Lee, R.D., Carter, L.R.: Modelling and forecasting U.S. mortality. J. Am. Stat. Assoc. **87**, 659–671 (1992)
9. Renshaw, A.E., Haberman, S.: Lee-Carter mortality forecasting: a parallel generalised linear modelling approach for England and Wales mortality projections. Appl. Stat. **52**, 119–137 (2003)
10. Renshaw, A.E., Haberman, S.: On the forecasting of mortality reduction factors. Insur. Math. Econ. **32**, 379–401 (2003)
11. Renshaw, A.E., Haberman, S.: Lee-Carter mortality forecasting with age specific enhancement. Insur. Math. Econ. **33**, 255–272 (2003)
12. Renshaw, A.E., Haberman, S.: A cohort-based extension to the Lee-Carter model for mortality reduction factors. Insur. Math. Econ. **38**(3), 556–570 (2006)

# Combinatorial Nonlinear Goal Programming for ESG Portfolio Optimization and Dynamic Hedge Management

**Gordon H. Dash Jr. and Nina Kajiji**

**Abstract** Compared to their fundamentally weighted counterparts naively diversified investment portfolios that embrace environmental, sustainability and governance (ESG) factors are known to experience enhanced long-term investment performance. This paper introduces a combinatorial nonlinear multiple objective optimization model to diversify the short-term ESG portfolio. The expectation of long-term wealth creation from an ESG portfolio is also examined. This latter investment objective is explored by implementing a discrete period ESG portfolio re-balancing with attached dynamic hedging. Post simulation, we report comparatively higher Sharpe ratios and lower VaR metrics for the multiobjective and dynamically hedged ESG portfolio investment style.

**Keywords** Combinatorial goal programing · ESG-factor portfolios · Hedging

## 1 Introduction

The accessibility of data on firm-wide ESG-factors presents investors with the opportunity to refine their portfolio diversification goals to better achieve long-term value creation. The ESG investment approach offers a direct benefit to non-profit organizations. Largely, non-profit organizations exist to offer sustainable social and environmental value to the public. This goal is clearly consistent with the objectives of the ESG-factor approach to investing and long-term value creation. As a local non-profit organization, Girl Scouts of Rhode Island (GSRI) derives its long-term

G.H. Dash Jr. (✉)
Finance and Decision Sciences Area, College of Business Administration, University of Rhode Island, Kingston, RI, USA
e-mail: ghdash@uri.edu

N. Kajiji
Department of Computer Science and Statistics, University of Rhode Island, Kingston, RI, USA

N. Kajiji
The NKD-Group, Inc., Providence, RI, USA
e-mail: nina@nkd-group.com

C. Perna, M. Sibillo (eds.), *Mathematical and Statistical Methods for Actuarial Sciences and Finance*, DOI 10.1007/978-3-319-05014-0_18,

view on service delivery and its adoption of sustainable investing (SI) from the philanthropic guidelines of its global parent, the World Association of Girl Guides and Girl Scouts.

The purpose of this research is to demonstrate how the portfolio investment decision faced by GSRI can be specified as a short-term ESG-factor based combinatorial nonlinear multiple objective optimization (MINLGP) model. We also examine the characteristics of long-term risk mitigation by subjecting the short-term MINLGP portfolio to an optimally derived dynamic futures hedge.

## 2 Literature Review

Contemporary research reports that companies with high ESG scores tend to have less company-specific risk ([1, 7]). A broader examination of extant research shows companies with an adherence to ESG criteria contribute significantly to the overall reduction in portfolio risk ([5, 6]). However, for many non-profit organizations, state investment authorities and sovereign wealth management firms, the behavioral bias inherent in SI strategies presents a key barrier to practical implementation. This deterrent raises important questions about the steps these professional fund managers can take to gain the long-term benefits offered by SI and ESG-factor investing. An informed response requires institutional investment policymakers to: (a) adopt a greater breadth in the development of an approved investment list, and (b) increase the scope of investment policy in a manner that unites short-term risk mitigation with a long-term value creation model—a model that embraces factors beyond the traditional financial valuation method. Urwin [9] argues that his five-factor approach to SI produces two important advantages: (a) lower operational costs, and (b) an investor's objective function that is more aligned with (long-term) inter-generational equity management. The precepts in [9] are consistent with the MINLGP specification proposed in this research.

## 3 Combinatorial Nonlinear Goal Programming

Under the SI approach to diversification it is not always possible to satisfy all investment objectives simultaneously. For this reason alone the modeling process preferred for SI factor diversification conforms to recent advances in convex nonlinear multiple objective modeling (for a review, see [3, 8] and [4]). Modeling the Sharpe portfolio diversification problem by MINLGP is a three-step application. The steps are: (1) after applying negative screens, define the set of investable securities; (2) add the goal-directed constraint set to the MINLGP model; and (3) choose an action for the futures hedge, $F$.

The completed MINLGP is stated as:

$$MINLGP = Min\, Z = \left[P_1\left(h^-, h^+\right), P_2\left(h^-, h^+\right), \ldots, P_L\left(h^-, h^+\right)\right]$$
$$\text{S.T.} \quad Ax + Bf + h^- - h^+ = b.$$

Where: $x, f, h^-, h^+ \geq 0$, $f \in Z^F$, and, $x \in \Re^{n-F}$.

Here $m$ is the number of goal constraints such that $A \in \Re^{mxn}$, $B \in Z^{mxF}$, $b \in \Re^m$, and $Z$ quantifies the attainment of $L$ hierarchical levels such that $P_1(h^-, h^+) > P_2(h^-, h^+) > \cdots > P_L(h^-, h^+)$. When necessary, scaling effects applied to $P_l(h^-, h^+)$ are defined by the nature and numerical definition of the separable goal programming model. We note that in the absence of hedging constraints (i.e., $F = 0$) we obtain the solution to the unconstrained convex Sharpe goal program. Under this specification $b$ is the $m$-component vector of goal targets while $h^-$ and $h^+$ are $m$-component column vectors that capture goal under- and over-achievement, respectively. Lastly, we define the optimal solution to the convex MINLGP, $x^*$, as the one that satisfies all hierarchical levels as much as possible. The convexity property of the MINLGP permits adding the necessary *if-then* constraint to control the dynamic trading of the contingent claim contract. Throughout, we refer to the formulation where $F > 0$ as the "dynamically hedged Sharpe MINLGP optimization."

## 4 Application and Results

In this section of the paper we model the short-term ESG diversification problem as a single-period Sharpe efficient portfolio with $N_f$ contingent claims attached. The model is applied after obtaining historical daily price data for $n$ instruments, $n \in \{1, \ldots, 75\}$, and the market proxy over the period from November 1, 2012 through November 31, 2012, inclusive. Next we formulate relevant goal constraints.

The Sharpe MINLGP approach to mean-variance minimization relies upon hierarchical goal equations to control efficient asset allocation per unit of systematic portfolio risk (for details see [4]). The model proposed in this research augments the basic Sharpe MINLGP objective function to include ESG factors. Additionally, integrability relations are added to the canonical specification in order to open the dynamic hedge with an optimal number of futures contracts ($N_f$). For post simulation comparative analytics we follow [2]. Here we compare the long-term risk-return differences between ESG and naively diversified portfolios. For a designated return level, preliminary results report a small dollar loss of approximately \$14,000 for the unhedged MINLGP GSRI portfolio. The equivalent naively diversified portfolio experience a loss of just under \$20,000. As expected, interim period hedging produced wealth gains that more than offset reported investment losses. The gains from dynamically hedging the GSRI portfolio totaled \$34,299; an outcome that produced lower VaR metrics and higher Sharpe ratios across the simulated time periods.

## 5 Summary and Conclusions

The research presented in this paper was inspired by non-profit investors who seek long-term consistency in risk-adjusted performance. With a clear view of the philanthropic role that guides non-profit decision-making, this research identified a feasible investment guide for these fund managers. By relying on ESG-factor optimized and dynamic interim period hedging, the research demonstrated a clear risk-mitigated advantage for the non-profit GSRI. Except for low rate of return portfolios, the results of applying the short-term MINLGP model across consecutive re-balancing periods in conjunction with interim period hedging produced superior risk-adjusted performance metrics. These results amplify the usefulness of interim period hedging in ESG-factor based sustainable investing.

## References

1. Bouslah, K., Kryzanowski, L., M'Zali, B.: Relationship between firm risk and individual dimensions of social performance. In: Proceedings of the Annual Conference of the Administrative Science Association of Canada, Montreal (Canada), vol. 32, pp. 105–122 (2011)
2. Daryl, C.R.M., Shawn, L.K.J.: The attenuation of idiosyncratic risk under alternative portfolio weighting strategies: recent evidence from the UK equity market. Int. J. Econ. Finance **4**, 1–14 (2012)
3. Dash, G.H. Jr., Kajiji, N.: A nonlinear goal programming model for efficient asset-liability management of property-liability insurers. Inf. Syst. Oper. Res. **43**(2), 135–156 (2005)
4. Dash, G.H. Jr., Kajiji, N.: On mulitobjective combinatorial optimization and dynamic interim hedging of efficient portfolios. Int. Trans. Oper. Res. (2014). doi:10.1111/itor.12067
5. Hoepner, A.G.F.: Portfolio diversification and environmental, social or governance criteria: must responsible investments really be poorly diversified? Available at http://ssrn.com/abstract=1599334 (2010)
6. Hoepner, A.G.F., Rezec, M., Siegl, K.S.: Does pension funds' fiduciary duty prohibit the integration of environmental responsibility criteria in investment processes? A realistic prudent investment test. Available at http://ssrn.com/abstract=1930189 (2013)
7. Oikonomou, I., Brooks, C., Pavelin, S.: The impact of corporate social performance on financial risk and utility: a longitudinal analysis. Financ. Manag. **41**, 483–515 (2012)
8. Pennanen, T.: Introduction to convex optimization in financial markets. Math. Program. **134**(1), 91–110 (2012)
9. Urwin, R.: Allocations to sustainable investing. Towers Watson Technical Paper No. 1656955. Available at http://ssrn.com/abstract=1656955 (2010)

# On the Geometric Brownian Motion with Alternating Trend

**Antonio Di Crescenzo, Barbara Martinucci, and Shelemyahu Zacks**

**Abstract** A basic model in mathematical finance theory is the celebrated geometric Brownian motion. Moreover, the geometric telegraph process is a simpler model to describe the alternating dynamics of the price of risky assets. In this note we consider a more general stochastic process that combines the characteristics of such two models. Precisely, we deal with a geometric Brownian motion with alternating trend. It is defined as the exponential of a standard Brownian motion whose drift alternates randomly between a positive and a negative value according to a generalized telegraph process. We express the probability law of this process as a suitable mixture of Gaussian densities, where the weighting measure is the probability law of the occupation time of the underlying telegraph process.

**Keywords** Alternating counting process · Exponential random times · Occupation time · Telegraph process

## 1 Introduction

A customary assumption in mathematical finance theory is that the price of a risky asset evolves according to the time-homogeneous geometric Brownian motion. Various other models can be employed in order to capture the alternation between increasing and decreasing trends observed often in true markets. A basic model characterized by alternating trend is the telegraph process. This is used to describe the random motion of a particle that runs with finite speed on the real line and alternates between two possible directions of motion at random time instants driven by a

A. Di Crescenzo (✉) · B. Martinucci
Department of Mathematics, University of Salerno, Fisciano (SA), Italy
e-mail: adicrescenzo@unisa.it

B. Martinucci
e-mail: bmartinucci@unisa.it

S. Zacks
Department of Mathematical Sciences, Binghamton University, Binghamton, NY, USA
e-mail: shelly@math.binghamton.edu

C. Perna, M. Sibillo (eds.), *Mathematical and Statistical Methods for Actuarial Sciences and Finance*, DOI 10.1007/978-3-319-05014-0_19,

homogeneous Poisson process. See [7] for the explicit form of the probability law, which is governed by the telegraph equation, and for various other results on such random process. We refer to the book [6] for a comprehensive treatise on this topic.

Among the several generalizations of the telegraph process considered in the literature we recall the geometric telegraph process. It was proposed in [2] as a model in mathematical finance. Such a process has bounded variations and exhibits increasing and decreasing trends alternating according to a suitable counting process. It was improved in [8] and [9] by inclusion of jumps, which avoid arbitrage opportunities thanks to the martingale property.

Aiming to deal with a more flexible stochastic process of interest in financial modeling, in this note we investigate a geometric Brownian motion with alternating trend. This is defined as the exponential of a Brownian motion whose drift alternates randomly between a positive and a negative value (see [3]), according to a generalized telegraph process. In Sect. 2 we provide the probability law of this process, expressed as a suitable mixture. As example, we consider two cases when the times separating consecutive trend changes have exponential distribution with constant and linear rates, the latter case being stimulated by studies in [1] and [5].

## 2 The Model and Its Probability Law

We assume that the price of a risky asset is described by the stochastic process

$$S(t) = s_0 \exp\big[Y(t) + \sigma B(t)\big], \quad t \geq 0, \tag{1}$$

with $\sigma > 0$, and where $\{Y(t)\}$ is a generalized (integrated) telegraph process, and $\{B(t)\}$ is a standard Brownian motion. Let $Y(t)$ and $B(t)$ be independent processes, and let $Y(0) = B(0) = 0$, so that $S(0) = s_0 > 0$ is the price at time 0. The telegraph process $Y(t)$ describes the alternating trend of $S(t)$, with velocities $c > 0$ and $-v < 0$, which alternate in time according to an independent alternating counting process $N(t)$ governed by independent random times $\{U_1, U_2, \ldots\}$ and $\{D_1, D_2, \ldots\}$. The random time $U_k$ (resp. $D_k$) describes the $k$-th random period during which $Y(t)$ has velocity $c$ (resp. $-v$). We denote by $V(t)$ the velocity of $Y(t)$ at time $t \geq 0$, and assume that initially it is equal to $c$ or $-v$ with equal probability, i.e. $P\{V(0) = c\} = P\{V(0) = -v\} = \frac{1}{2}$, with $V(0)$ independent from $B(t)$. Thus we have:

$$Y(t) = \int_0^t V(s)\,\mathrm{d}s, \quad V(t) = \frac{c-v}{2} + \mathrm{sgn}\big(V(0)\big)\frac{c+v}{2}(-1)^{N(t)}, \ t > 0; \tag{2}$$

see [3] for other details. In particular, the probability law of $S(t)$ can be expressed in terms of the distribution of the following occupation time:

$$W(t) := \int_0^t \mathbf{1}_{\{V(s)=c\}}\mathrm{d}s, \quad t > 0, \tag{3}$$

which is the fraction of time during which $Y(t)$ had velocity $c$ in $[0, t]$. Set $\overline{F}_{D_1}(t) = P(D_1 > t)$ and $\overline{F}_{U_1}(t) = P(U_1 > t)$. Moreover let $f_U^{(n)}$ and $F_U^{(n)}$ (resp.

$f_D^{(n)}$ and $F_D^{(n)}$) be respectively the density and the distribution function of $U^{(n)} := U_1 + \cdots + U_n$ (resp. $D^{(n)} := D_1 + \cdots + D_n$). Recalling Theorem 3.1 of [3], the probability law of $W(t)$ has a discrete mass on points 0 and $t$ such that

$$P\big[W(t)=0\big] = \frac{1}{2}\overline{F}_{D_1}(t), \qquad P\big[W(t)=t\big] = \frac{1}{2}\overline{F}_{U_1}(t), \tag{4}$$

and a density $\psi(w,t) := \frac{\partial}{\partial w}P[W(t) \le w]$ expressed in series form as

$$\begin{aligned}\psi(w,t) &= \sum_{n=1}^{+\infty}\big[F_U^{(n)}(w) - F_U^{(n+1)}(w)\big] f_D^{(n)}(t-w) \\ &\quad + \sum_{n=0}^{+\infty}\big[F_D^{(n)}(t-w) - F_D^{(n+1)}(t-w)\big] f_U^{(n+1)}(w) \\ &\quad + \sum_{n=0}^{+\infty}\big[F_U^{(n)}(w) - F_U^{(n+1)}(w)\big] f_D^{(n+1)}(t-w) \\ &\quad + \sum_{n=1}^{+\infty}\big[F_D^{(n)}(t-w) - F_D^{(n+1)}(t-w)\big] f_U^{(n)}(w), \quad 0 < w < t. \end{aligned} \tag{5}$$

We are now able to evaluate the distribution function of process $S(t)$. Recall that Eqs. (2) and (3) yield $Y(t) = (c+v)W(t) - vt$, for $t > 0$. Hence, making use of (1) and conditioning on $W(t)$ we have, for $x > 0$ and $t > 0$,

$$P\big[S(t) \le x\big] = \int_0^t \Phi\left(\frac{1}{\sigma}\left[\ln\frac{x}{s_0} - (c+v)w + vt\right]\right) \mathrm{d}F_{W(t)}(w),$$

where as usual $\Phi(\cdot)$ denotes the standard normal distribution. By differentiation and recalling Eqs. (4) and (5) we obtain the probability density of $S(t)$, say $p(x,t)$, given by the following mixture of the standard normal density $\phi(\cdot)$, for $x > 0$ and $t > 0$:

$$\begin{aligned} p(x,t) &= \frac{1}{2}\overline{F}_{D_1}(t)\phi\left(\frac{1}{\sigma}\left(\ln\frac{x}{s_0} + vt\right)\right) + \frac{1}{2}\overline{F}_{U_1}(t)\phi\left(\frac{1}{\sigma}\left(\ln\frac{x}{s_0} - ct\right)\right) \\ &\quad + \frac{1}{\sigma x}\int_0^t \psi(w,t)\phi\left(\frac{1}{\sigma}\left[\ln\frac{x}{s_0} - (c+v)w + vt\right]\right)\mathrm{d}w. \end{aligned} \tag{6}$$

As a case study we assume that the random times $U_k$ and $D_k$, $k = 1, 2, \ldots$, have exponential distribution with parameters $\lambda_k$ and $\mu_k$, respectively, and consider the two instances with (i) constant parameters $\lambda_k = \lambda$ and $\mu_k = \mu$, and (ii) linear parameters $\lambda_k = \lambda k$ and $\mu_k = \mu k$. See Propositions 5.1 and 5.4 of [3], where the density $\psi(w,t)$ has been evaluated in such two cases. This allows to apply a numerical procedure able to obtain some plots of the density (6) given in Fig. 1.

We remark that a simulation-based approach useful to face with the first-passage time problem through constant boundaries for $S(t)$ is provided in [4].

In conclusion, we point out that (1) can be viewed as a starting point for more tight models. Indeed, the problem of existence of arbitrage opportunities for (1) will

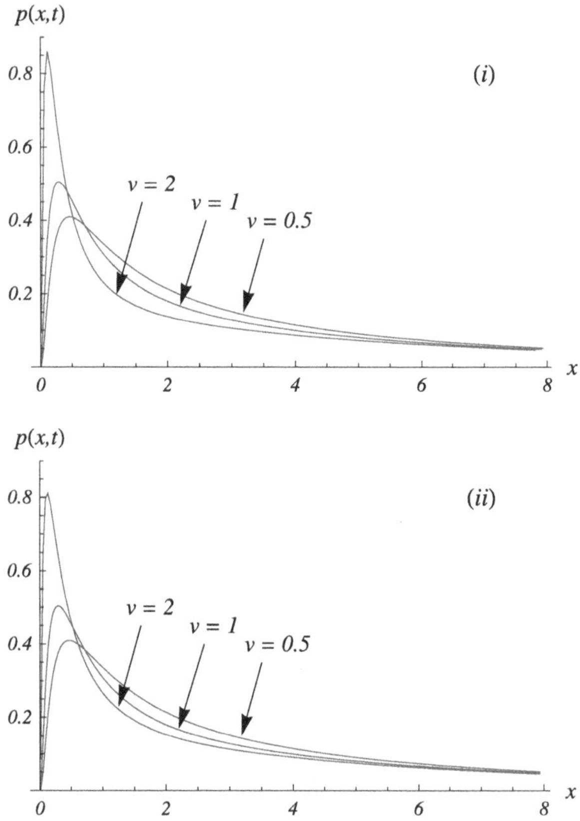


**Fig. 1** Density (6) for $t = 1$, $\sigma = 1$, $s_0 = 1$, $c = 1$, for exponential times $U_k$ and $D_k$ with (**i**) constant rates and (**ii**) linear rates, with $\lambda = \mu = 1$

be the object of a future investigation, aimed to prove that $S(t)$ is a martingale under specific assumptions, for instance modifying the process by inclusion of jumps.

## References


1. Di Crescenzo, A., Martinucci, B.: A damped telegraph random process with logistic stationary distribution. J. Appl. Probab. **47**, 84–96 (2010)
2. Di Crescenzo, A., Pellerey, F.: On prices' evolutions based on geometric telegrapher's process. Appl. Stoch. Models Bus. Ind. **18**, 171–184 (2002)
3. Di Crescenzo, A., Zacks, S.: Probability law and flow function of Brownian motion driven by a generalized telegraph process. Methodol. Comput. Appl. Probab. (2013). doi:10.1007/s11009-013-9392-1
4. Di Crescenzo, A., Di Nardo, E., Ricciardi, L.M.: Simulation of first-passage times for alternating Brownian motions. Methodol. Comput. Appl. Probab. **7**, 161–181 (2005)
5. Di Crescenzo, A., Martinucci, B., Zacks, S.: On the damped geometric telegrapher's process. In: Perna, C., Sibillo, M. (eds.) Mathematical and Statistical Methods for Actuarial Sciences and Finance, pp. 175–182. Springer, Berlin (2012). ISBN 978-88-470-2341-3
6. Kolesnik, A.D., Ratanov, N.: Telegraph Processes and Option Pricing. Springer, Berlin (2013). ISBN 978-3-642-40525-9

7. Orsingher, E.: Probability law, flow function, maximum distribution of wave-governed random motions and their connections with Kirchoff's laws. Stoch. Process. Appl. **34**, 49–66 (1990)
8. Ratanov, N.: A jump telegraph model for option pricing. Quant. Finance **7**, 575–583 (2007)
9. Ratanov, N.: Option pricing model based on a Markov-modulated diffusion with jumps. Braz. J. Probab. Stat. **24**, 413–431 (2010)

# Empirical Evidences on Predictive Accuracy of Survival Models

**Emilia Di Lorenzo, Michele La Rocca, Albina Orlando, Cira Perna, and Marilena Sibillo**

**Abstract** The paper focuses on a stochastic proportional hazard model depicting the evolution of the force of mortality; in particular the real data are plotted against a specific survival model by means of the stochastic process that describes their ratio. The predictive accuracy of the survival model is investigated, since, by means of the calibrated "ratio process", its forecasting skills are assessed. A statistical analysis is developed in order to test the capacity the assumed survival model has to follow the real behavior of the observed data.

**Keywords** Survival models · Longevity risk · Parametric bootstrap

## 1 Introduction

The mathematical models describing the evolution of survival still leave many open problems, being hard a good fitting of the real data. By virtue of the socio-economic importance of the phenomenon they represent, the issue is pregnant in several prac-

E. Di Lorenzo (✉)
Dept. of Economic and Statistical Sciences, University of Naples Federico II, Napoli, Italy
e-mail: emilia.dilorenzo@unina.it

M. La Rocca · C. Perna · M. Sibillo
Dept. of Economics and Statistics, University of Salerno, Fisciano (SA), Italy

M. La Rocca
e-mail: larocca@unisa.it

C. Perna
e-mail: perna@unisa.it

M. Sibillo
e-mail: msibillo@unisa.it

A. Orlando
Istituto per le Applicazioni del Calcolo, CNR, Napoli, Italy
e-mail: a.orlando@na.iac.cnr.it

C. Perna, M. Sibillo (eds.), *Mathematical and Statistical Methods for Actuarial Sciences and Finance*, DOI 10.1007/978-3-319-05014-0_20,

tical applications. In fact, the recent decades are witnessing a continuous decline in the mortality rate and, simultaneously, a marked phenomenon of rectangularization and expansion of the survival function. Within the socio-economic development, the current guidelines aim at exploiting the survival trend as competitive leverage, and, at the same time, they address to protect the rights of the elderly, as consumers. In this context, it is necessary to frame a number of financial instruments, particularly with regard to the pension sector. The aim of the paper is in line with the need to understand if existing survival models are able to represent the real behavior of the survival phenomenon. In the paper we consider the stochastic proportional hazard model proposed in [1] for the evolution of the force of mortality; in particular the real data are plotted against a specific survival model by means of the stochastic process that describes their ratio. We model the ratio by means of a CIR stochastic process and following this guideline, the predictive accuracy of the survival model can be developed, since, by means of the calibrated ratio process, we can assess its forecasting skills. The idea, with strong applicative impact, is to pick out where, in what age interval, for what population characteristics and so on, the predictive capacity of the chosen survival model can be judged good.

## 2 The Model

Let us consider a person aged $x$ (say $(x)$). According to the model proposed by Di Lorenzo et al. in 2006 (cf. [1]), we consider a representation of the mortality rates based on a stochastic model apt to involve properly the observed data. In particular, we consider that the force of mortality dynamic follows a stochastic process obtained by modifying the deterministic anticipated realizations of the force of mortality by means of a stochastic process.

Let $\mu_{x+t}$ the survival model chosen to describe the future behavior of the mortality rates capturing the survival improvement due to longevity. The idea is to deep the capability the process $\mu_{x+t}$ has in repeating the observed data. The evolution in time of the real survival phenomenon (observed data) is described by the following model:

$$B_{x,t} = \mu_{x+t} Y_t, \tag{1}$$

where $Y_t$ is the process involved by the stochastic differential equation:

$$dY_t = \alpha(\gamma - Y_t)dt + \sigma\sqrt{Y_t}dW_t \tag{2}$$

$\alpha$, $\gamma$ and $\sigma$ being positive constants and $W_t$ a Wiener process. $Y_t$ is a continuous and positive process, which doesn't reach 0 for $2\alpha \geq \sigma^2$. As remarked in [1], a good choice of the deterministic function determines the reversion towards the long term value of the process with $\gamma = 1$; in this sense the long term value coincides with the position of the process in $t = 0$, just when its value corresponds to the observation at the beginning. In that case $B_{x,t} = \mu_{x+t}$ and $Y_0 = 1$.

The study focuses on the stochastic ratio $Y_t = \frac{B_{x,t}}{\mu_{x+t}}$ and develops a statistical analysis in order to test the capacity the survival function assumed to model $\mu_{x+t}$

has in following the real behavior of the observed data, in particular for the cohort of interest. The final goal aims at determining, for a specific population cohort of interest, a proper calibration procedure, which allows, on the basis of (2), to forecast accurate mortality/survival projections.

## 3 Parameter Estimation and Bootstrap Inference

Let $\{Y_t\}_{t=0}^n$ be observations from process $Y_t$. Then they satisfy the following discrete time series model

$$Y_t = e^{-\alpha\delta} Y_{t-1} + \gamma\left(1 - e^{-\alpha\delta}\right) + \varepsilon_t$$

where $E(\varepsilon_t) = 0$, $E(\varepsilon_t \varepsilon_s) = 0\ \forall t \neq s$ and $E(\varepsilon_t^2) = \frac{1}{2}\sigma^2\alpha^{-1}(1 - e^{-2\alpha\delta})Y_{t-1}$.

Assuming that $\varepsilon_t$ is Gaussian distributed, pseudo-MLEs can be obtained as (cf. [4]):

$$\hat{\alpha} = -\delta^{-1}\log(\hat{\theta}_1) \qquad \hat{\gamma} = \hat{\theta}_2 \qquad \hat{\sigma}^2 = \frac{2\hat{\alpha}\hat{\theta}_3}{1 - \hat{\theta}_1^2} \tag{3}$$

where

$$\hat{\theta}_1 = \frac{n^{-2}\sum_{t=1}^n Y_t \sum_{t=1}^n Y_{t-1}^{-1} - n^{-1}\sum_{t=1}^n Y_t Y_{t-1}^{-1}}{n^{-2}\sum_{t=1}^n Y_{t-1}\sum_{t=1}^n Y_{t-1}^{-1} - 1};$$

$$\hat{\theta}_2 = \frac{n^{-1}\sum_{t=1}^n Y_t Y_{t-1}^{-1} - \hat{\theta}_1}{(1-\hat{\theta}_1)n^{-1}\sum_{t=1}^n Y_{t-1}^{-1}}$$

$$\hat{\theta}_3 = n^{-1}\sum_{t=1}^n \left\{Y_t - Y_{t-1}\hat{\theta}_1 - \hat{\theta}_2(1-\hat{\theta}_1)\right\}^2 Y_{t-1}^{-1}.$$

Inference on the unknown parameters $\alpha$, $\gamma$ and $\sigma^2$ can be gained by using a parametric bootstrap scheme.

The empirical distribution functions of these bootstrap replicates can be used to make an inference on the unknown parameters $\alpha$, $\gamma$ and $\sigma^2$. Bootstrap confidence intervals with nominal level $1-\alpha$ can be easily obtained by taking as confidence limits the percentiles of order $\alpha/2$ and $1-\alpha/2$ of the bootstrap distributions.

## 4 Results and Final Remarks

The model presented in Sect. 2 has been implemented choosing $\mu_{x+t}$ modeled by the Lee Carter stochastic survival function, as in Lee and Carter (cf. [3]). The analysis has been performed on the French total population mortality data got by Human Mortality database website (cf. [2]). In particular, the Lee Carter parameter estimation has been based on data from 1861 to 1960. The statistical tests on the model

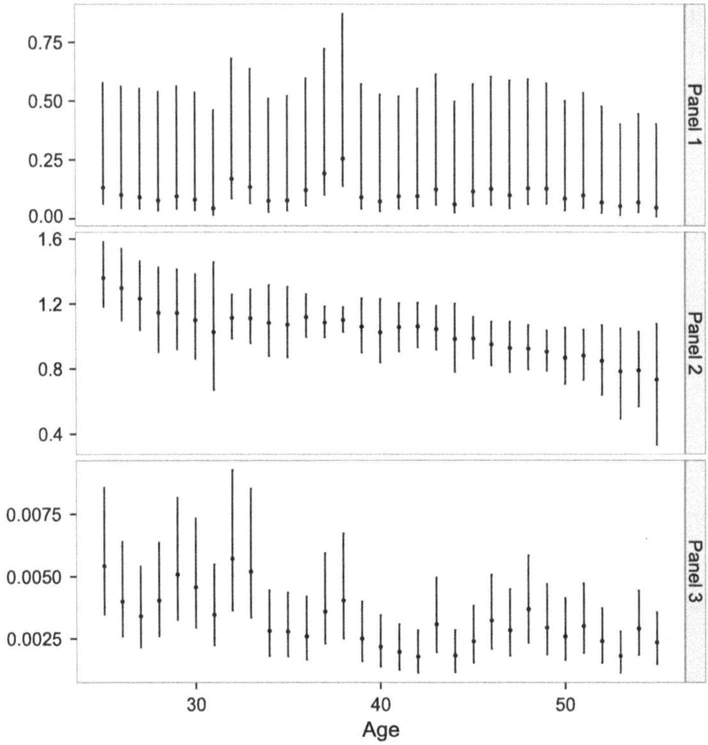


**Fig. 1** Bootstrap percentile confidence intervals, with nominal confidence level equal to 0.95, for the parameters of CIR models estimated for ages ranging from 25 to 55. Panels 1, 2 and 3 refer, respectively, to parameters $\alpha$, $\gamma$ and $\sigma^2$

in (1) have been carried out in the period 1961–2007. Here the inference procedure on the CIR model has been executed for the age range 25–55, considering the practical motive of this application referred to post retirement insurance covering purposes. As clearly showed in Panel 2 of Fig. 1, at 0.95 confidence level the intervals around 1 are sufficiently small in the age interval of interest.

## References

1. Di Lorenzo, E., Sibillo, M., Tessitore, G.: A stochastic proportional hazard model for the force of mortality. JoF 529–553 (2006)
2. Human Life-Table Database. Max Planck Institute for Demographic Research, Rostock (Germany), Department of Demography at the University of California, Berkeley (USA), and the Institut National d'Étude Démographiques, Paris (2013). http://www.lifetable.de
3. Lee, R.D., Carter, L.R.: Modeling and forecasting US mortality. J. Am. Stat. Assoc. **87**(419), 659–671 (1992)
4. Tanga, C.Y., Chen, S.X.: Parameter estimation and bias correction for diffusion processes. J. Econom. **149**, 65–81 (2009)

# RedES™, a Risk Measure in a Pareto-Lévy Stable Framework with Clustering

**Riccardo Donati and Marco Corazza**

**Abstract** In this communication: (1) we present RedES™, an Expected Shortfall based risk measure developed in the framework of the Pareto-Lévy stable distributions with clustering; (2) we apply it to about 3,000 equity stocks. The results show that RedES™ is able to take into account the fat tail effects in a robust manner.

**Keywords** RedES™ · Pareto-Lévy stable distributions · Clustering · Equity stocks

## 1 Introduction

The shape of log-return distributions is crucial for many financial applications, from risk management to portfolio optimization. In order to detect the "right" distribution class, instead of performing several empirical analyses, we assume a simple time-invariance hypothesis, as Mandelbrot did in the 60s of the past century (see, for instance, [1]). This hypothesis straightly leads to the Pareto-Lévy stable (PLs) distributions class.

In this communication we first present RedES™, an Expected Shortfall (ES) based risk measure developed in the framework of the PLs distributions with clustering. Then we apply it to about 3,000 equity stocks.

In short, a distribution is PLs if its form is invariant under addition. The parameterization of the class involves four quantities: $\alpha$, the stability parameter; $\beta$, the skewness parameter; $\mu$, the location parameter; $\sigma$, the scale parameter (for PLs-based models in finance see [2]). Generally, the probability density function (PDF) of a PLs distribution is not analytically expressible, resulting in some computational difficulties that have to be carefully managed. Another tricky aspect arises from infinite variance.

R. Donati
Redexe S.r.l., Piazzale Giusti 8, 36100 Vicenza, Italy
e-mail: r.donati@redexe.net

M. Corazza (✉)
Department of Economics, Ca' Foscari University of Venice, Sestiere di Cannaregio 873, 30121 Venezia, Italy
e-mail: corazza@unive.it

C. Perna, M. Sibillo (eds.), *Mathematical and Statistical Methods for Actuarial Sciences and Finance*, DOI 10.1007/978-3-319-05014-0_21,

## 2 RedES™ and Its Applications

The construction and the utilization of RedES™ can be itemized as follows:

- First, we test the ability of PLs distributions to fit real log-returns. To this end we consider the constituents of the NYSE Composite, a stock market index covering all common stocks listed on the New York Stock Exchange. More than 2,000 stocks constitute this index, about 1,600 from United States (US) companies and about 400 from non-US companies. Further, we consider the constituents of the NASDAQ Composite, a stock market index of the common stocks and similar securities on the NASDAQ stock market. It has more than 3,000 components. Also in this stock market both US and non-US companies are listed. Our tests focuses on two time frames: 20 years (07/1991 to 07/2011) for 795 stocks, and 10 years (07/2001 to 07/2011) for 2,233 stocks.

  For the considered constituents, we first calculate the daily log-return (plus dividend yield) time series, the normal best fit, and the PLs best fit using the maximum likelihood method. Then, we calculate the $p$-value statistic and we perform the $\chi^2$ test. Finally, we study the quantity $g = -\mathrm{Log}_{10}(p\text{-value})$. An high $g$ value suggests that it is unlikely that the data comes from the considered theoretical distribution.

  In Fig. 1 we graphically report the results for both the time frames. The values of $g$ associated to the PLs best fit is generally lower than the one associated to the normal best fit, especially in the tails.
- Then, we investigate the stability parameter $\alpha$ relating it to the industry sector and, through a Kruskal-Wallis test used as a metric, we cluster all the involved industry sectors. Four clusters with similar $\alpha$ result. Given this, we build RedES™ as follows: we calculate the log-returns of the investigated time series; for each of these time series we calculate the best PLs fit with $\alpha$ constrained to its cluster and we calculate the ES 99 % over that fit.
- Finally, in order to be able to homogeneously compare different risk magnitudes, we defined the risk class function $rc(r) = \log_2(100|r| + 1)$, where $r$ is a given estimation of risk in terms of log-return distribution.

Note that, to the best of our knowledge, both the calculation of the best PLs fit with $\alpha$ constrained to its industrial sector cluster and the introduction of the risk class function $rc(\cdot)$ constitute elements of novelty.

Ended this phase, we deal with the following question: is RedES™ able to offer a reasonably "good" measure of risk even if only short time series are available?

In order to provide an answer, given that we consider as ex-ante "good" risk measure the StableES 99 % over 20 years, we calculate the errors introduced considering only a 2 year time window using the following estimators for all the considered stocks: NormalES 99 % 2 years; StableES 99 % 2 years; RedES™ 2 years. In Fig. 2 we report the PDFs of the differences between each of the later risk measures and the StableES 99 % over 20 years.

Notice that RedES™ 2 years is very close to StableES 99 % 20 years, unlike NormalES 99 % 2 years and StableES 99 % 2 years. So RedES™ appears appropriate for ex-ante calculating risk also when only short time series are available.

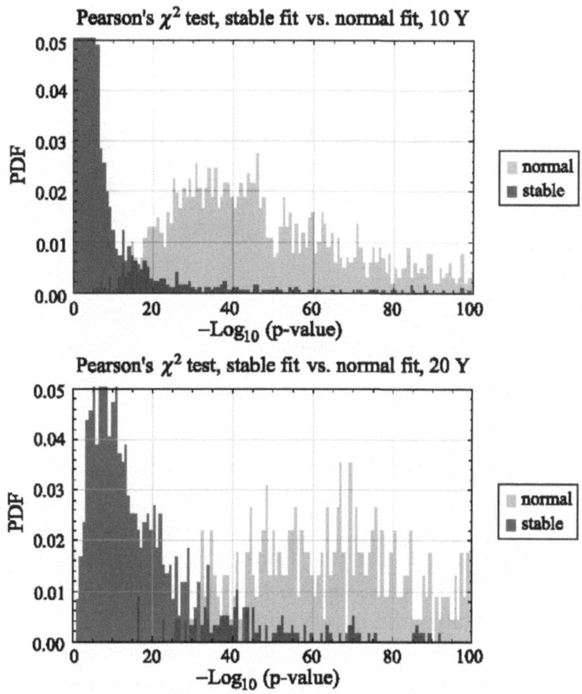


**Fig. 1** Behavior of the $\chi^2$ test for both the considered time frames

Finally, we consider the difference between the risk classes *rc*(RedES™ 2 years) and *rc*(StableES 99 % 20 years), relating such difference with respect to the market capitalization of the stocks. In Fig. 3 we graphically report the results.

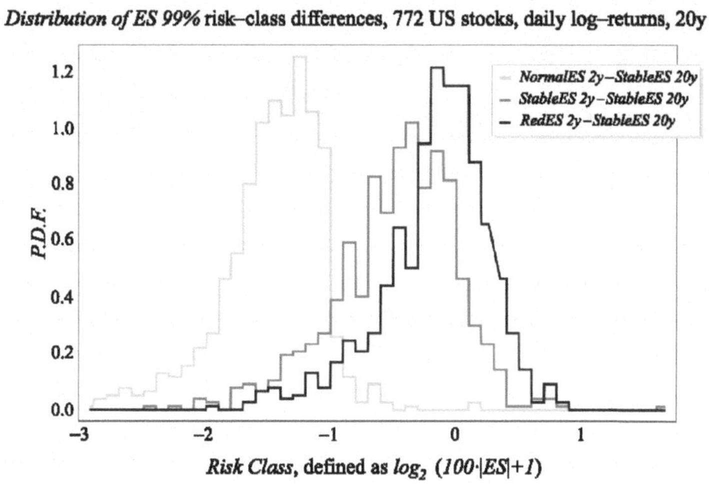


**Fig. 2** PDFs of: NormalES 99 % 2y–StableES 99 % 20y; StableES 99 % 2y–StableES 99 % 20y; RedES™ 2y–StableES 99 % 20y

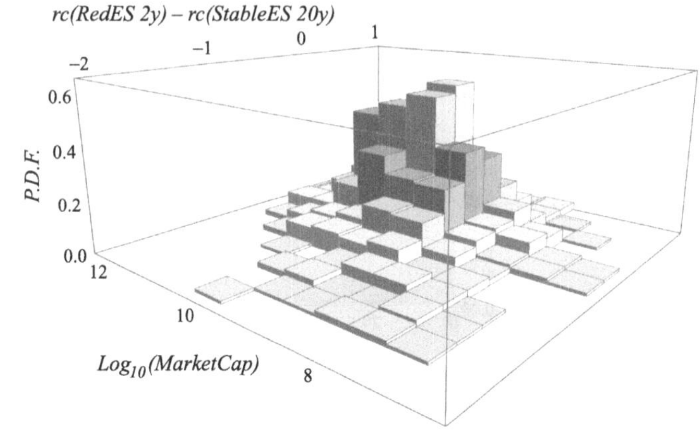


**Fig. 3** PDF of *rc*(RedES™ 2 years)–*rc*(Stable 99 % 20 years) with respect to the market capitalization of the stocks

It is possible to see that RedES™ 2 years is very close to the "true" risk StableES 99 % 20 years. Only a few shifts below −1 are observed essentially when market capitalization is very small.

Note that this robustness of RedES™ with respect to both the time window length and the market capitalization of the stocks constitute a further element of novelty of our approach.

## 3 Some Concluding Remarks

We can conclude that RedES™, a risk measure built in a Pareto-Lévy stable framework with clustering, is able: to correctly estimate risk taking into account the fat tail effects; to work well also when only short time series are available; to be robust with respect to the market capitalization of the stocks.

## References

1. Mandelbrot, B.B.: The variation of certain speculative prices. J. Bus. **36**(3), 394–419 (1963)
2. Rachev, S.T., Kim, Y.S., Bianchi, M.L., Fabozzi, F.J.: Financial Models with Lévy Processes and Volatility Clustering. Wiley, Hoboken (2011)

# Run-Off Error in the Outstanding Claims Reserves Evaluation

**Nicolino Ettore D'Ortona and Giuseppe Melisi**

**Abstract** The variability of claim costs represents an important risk component, which should be taken into account while implementing the internal models for solvency evaluation of an insurance undertaking. This component can generate differences between future payments for claims and the provisions set aside for the same claims (run-off error). If the liability concerning the claims reserve is evaluated using synthetic methods, than the run-off error depends on the statistical method adopted. This work focuses on measuring the run-off error with reference to claims reserves evaluation methods applied to simulated run-off matrices for the claims-settlement development. The results from the numerical implementations provide us with useful insights for a rational selection of the statistical-actuarial method for the claims reserve evaluation on an integrated risk management framework.

**Keywords** Run-off error · Outstanding claims reserves · Stochastic simulation

## 1 Outstanding Claims Reserve

The random claim settlement regarding the accident year $i$ $(i = 0, 1, \dots, t)$ is given by the sum of a random number of claims, each one subject to a single claim settlement. Since the settlement claimed for every accident usually requires two or more payments, which can take place during the accident year or the subsequent years, the aggregated claims cost for every accident year can be represented as follows:

$$\widetilde{X}(i) = \sum_{j=0}^{t} \widetilde{X}(i, j) \quad i = 0, 1, \dots, t \tag{1}$$

N.E. D'Ortona · G. Melisi (✉)
University of Sannio, via delle Puglie 82, Benevento, Italy
e-mail: gmelisi@luiss.it

N.E. D'Ortona
e-mail: dortona@unisannio.it

C. Perna, M. Sibillo (eds.), *Mathematical and Statistical Methods for Actuarial Sciences and Finance*, DOI 10.1007/978-3-319-05014-0_22,

where $\widetilde{X}(i,j)$ represents the amount paid for settlements regarding claims incurred during the accident year $i$ and settled after $j$ years; $t$ represents the maximum number of deferment years considered for the total settlement of a single claim.

The random amount required for future settlements regarding claims not yet settled or reported, for each accident year, is given by:

$$\widetilde{R}(i) = \sum_{j=t-i+1}^{t} \widetilde{X}(i,j) \quad i = 1, 2, \ldots, t \tag{2}$$

The aggregate amount required is then given by the sum: $\widetilde{R} = \sum_{i=1}^{t} \widetilde{R}(i)$.

## 2 Introduction to Statistical Methods

The statistical methods for the outstanding claims reserve evaluation consist in the formulation of a forecasted value of necessary reserve, based on a projected analysis of the data obtained by the examination of relevant time series. In other words, an evaluation method provides an estimator $\widehat{R} = f(\widetilde{K}_0, \widetilde{K}_1, \ldots, \widetilde{K}_t)$ of the expected value for the outstanding claims reserve,[1] which depends on the information at disposal $K = (\widetilde{K}_0, \widetilde{K}_1, \ldots, \widetilde{K}_t)$ for each accident year.

The difference between future payments for claims settlements and the amount of the outstanding claims reserve, evaluated using a specific estimator, gives us the run-off error. The run-off error for each accident year can be represented as follows:

$$\widetilde{e}(i) = \widehat{R}(i) - \widetilde{R}(i) = \sum_{j=t-i+1}^{t} \left[\widehat{X}(i,j) - \widetilde{X}(i,j)\right] \quad i = 1, 2, \ldots, t. \tag{3}$$

In practice, the run-off error can be measured only after the completion of the claims settlement process. In this work, we will quantify the run-off error, simulating the claims settlement process until we obtain all the members of the run-off error formula $\widetilde{e}(i)$, $i = 1, 2, \ldots, t$.

For this purpose, we represent the random settlements, in each cell of the run-off matrix, with the following (collective) model:

$$\widetilde{X}(i,j) = \sum_{k=0}^{\widetilde{N}(i,j)} \widetilde{Y}_k(i,j) \quad i, j = 0, 1, \ldots, t \tag{4}$$

where $\widetilde{N}(i,j)$ represents the total number of claims for the accident year $i$, settled during the development year $j$; $\widetilde{Y}_k(i,j)$ represents the random settlement for the claim $k$ incurred during the accident year $i$ and settled after $j$ years.

Between the multiple procedures for the evaluation of outstanding claims reserve proposed in literature, four of them were chosen for this work, considering their

[1] In general, for the distribution of the claims outstanding reserve, other than the expected value we can estimate moments of order higher than 1 or even particular quantiles.

widespread utilization in the professional environment: the Chain Ladder method ([3–5, 10]), the separation method (arithmetic and geometric) [9], the Fisher-Lange method and the Bornhuetter-Ferguson method [1].

## 3 The Results of the Comparative Analysis

For the simulation of the amounts $\widetilde{X}(i, j)$ we have considered four methods, which are distinguished for the development rule concerning the claims settlement inside the run-off triangle.

A comparative analysis was set up for the examination of the run-off error amplitude regarding each estimating method, considering different sets of parameters, which were recursively modified predicting: a different level of inflation, a higher volatility of the settlement amount, a higher volatility of the disturbing factors characterizing the settlement process, various temporal profiles for the claims development. For each set of parameters were generated 4.100 settlement matrices with each one of the following simulation techniques. The inferior triangle of the future settlements was obtained for each simulated matrix. Therefore, gap indicators between estimated reserves and effective (simulated) reserves were calculated.

**Method of random development factors (*Method 1*)** This method simulates the run-off matrix through the steps described in [6]. According to the mean square error criterion, the Fisher-Lange method presents a higher level of preferability. Analysing the single accident years, we deduce that the Chain-Ladder method provides a less biased estimator, with the relevant exception of the last accident year.

**Method of backward calculated random development factors (*Method 2*)** The method generates the run-off matrix through the steps described in [6]. The Fisher-Lange method estimator shows the lower mean square error for both the single accident year estimation and the whole portfolio estimation. The Chain-Ladder estimator results to be the less biased estimator and shows the lower expected percentage error.

**Method of single settlements (*Method 3*)** This simulative method is ascribable to the estimation models proposed by Stanard [8] and by Buhlmann, Schnierper and Straub [2]. The simulating technique appears to be rather coherent in structure, with the claims development model upon which is based the Fisher-Lange method. All methods provide estimators with high levels of correlation with the estimated reserve.

**Pentikainen-Rantala method (*Method 4*)** This method operates through the steps described in [7]. In this case, according to both the expected percentage error criterion and the dispersion criterion, the geometric separation method presents the higher level of preferability.

**Table 1** Main results of numerical simulations—Mean Percentage Error

| Method | Chain-Ladder | Arithmetic Separation | Geometric Separation | Fisher-Lange | Bornhuette-Ferguson |
|---|---|---|---|---|---|
| Method 1 | 0.32 % | 0.26 % | 0.64 % | 0.82 % | −7.32 % |
| Method 2 | 0.16 % | −0.37 % | −2.01 % | 1.15 % | −3.74 % |
| Method 3 | 0.10 % | −0.74 % | 0.69 % | −0.16 % | 0.49 % |
| Method 4 | 1.12 % | 0.07 % | −0.07 % | 2.16 % | −1.67 % |

Table 1 shows the mean percentage error obtained by dividing the mean error of each estimation method to the estimated value of the claims reserve. These statistics allow you to know the sign of the error, and then the tendency of the evaluation methods to overestimate or underestimate the value of the reserve.

The numerical implementation results point out the following:

- the estimating methods produce a lower run-off error if applied to a development matrix which, despite not respecting some of the probabilistic hypothesis of the method, provide a settlement distribution according to the mechanism considered by the estimating model;
- some of the estimating methods despite showing a minor distortion of the reserve estimation for the entire portfolio, result imprecise in the prediction of the run-off for single accident years;
- the preferability of the estimating methods did not show particular sensibility to the choice of numerical values attributed to the parameters.

## References

1. Bornhuetter, R.L., Ferguson, R.E.: The actuary and IBNR. Proc. Casualty Actuar. Soc. **LIX**, 181–195 (1972)
2. Buhlmann, H., Straub, E., Schnieper, R.: Claims reserves in casualty insurance based on a probabilistic model. Mitt. - Schweiz. Ver. Versicher.math. **80**(1), 21–45 (1980)
3. Institute of Actuaries: Claims Reserving Manual. Lond. (1997)
4. Mack, T.: Measuring the variability of Chain Ladder reserve estimates. In: Casualty Actuar. Soc. Forum Spring, pp. 101–182 (1994)
5. Mack, T.: The prediction error for Bornhuetter-Ferguson. ASTIN Bull. **38**, 87–103 (2008)
6. Narayan, P., Warthen, T.: A comparative study of the performance of loss reserving methods through simulation. In: Casualty Actuar. Soc. Forum Summer, vol. I, pp. 175–195 (1997)
7. Pentikainen, T., Rantala, J.: A simulation procedure for comparing different claims reserving methods. ASTIN Bull. **22**, 191–216 (1992)
8. Stanard, J.N.: A simulation test of prediction errors of loss reserve estimation techniques. Proc. Casualty Actuar. Soc. **LXXII**, 181–195 (1986)
9. Taylor, G.C.: Claims Reserving in Non-life Insurance. North-Holland, Amsterdam (1986)
10. Verra, R.J.: A Bayesian generalised linear model for the Bornhuetter-Ferguson method of claims reserving. N. Am. Actuar. J. **8**, 67–89 (2004)

# Trajectory Based Market Models. Arbitrage and Pricing Intervals

**Sebastian Ferrando, Alfredo Gonzalez, Ivan Degano, and Massoome Rahsepar**

**Abstract** The paper introduces general, discrete, non probabilistic models and a natural global minmax pricing rule that, for a given option, leads to a pricing interval. Conditions are described for the absence of arbitrage and a dynamic programming local minmax optimization is defined that evaluates the pricing interval bounds.

**Keywords** Minmax · Pricing intervals · Non probabilistic

## 1 Introduction

The market model introduced in [3], by Britten-Jones and Neuberger (BJ&N), incorporates several important market features: it reflects the discrete nature of financial transactions, it models the market in terms of observable trajectories and incorporates practical constraints such as jump sizes as well as methodological constraints in terms of the quadratic variation. The book treatment of this model in [6] also emphasizes the fundamental characteristics of the model's assumptions. Here we summarize new fundamental results providing arbitrage-free models as well as a general computational framework for a general class of models encompassing the BJ&N setup. Proofs and many other results are fully described in [4].

---

S. Ferrando (✉) · M. Rahsepar
Department of Mathematics, Ryerson University, Toronto, Ontario, Canada
e-mail: ferrando@ryerson.ca

M. Rahsepar
e-mail: a.rahsepar@hotmail.com

A. Gonzalez · I. Degano
Department of Mathematics, Mar del Plata National University, Mar del Plata, Argentina

A. Gonzalez
e-mail: algonzal@mdp.edu.ar

I. Degano
e-mail: ivanldegano@yahoo.com.ar

C. Perna, M. Sibillo (eds.), *Mathematical and Statistical Methods for Actuarial Sciences and Finance*, DOI 10.1007/978-3-319-05014-0_23,

The framework of the paper is a tuple $\mathcal{M} = (\mathcal{S}, \mathcal{H})$ where the set $\mathcal{S}$ is a set of discrete trajectories and $\mathcal{H}$ a collection of hedging/investment portfolios. For a given option $Z$, of American or European type, we describe a pricing interval $[\underline{V}(Z), \overline{V}(Z)]$ such that any market price falling outside of this interval generates an arbitrage opportunity relative to the market model $\mathcal{M}$. This arbitrage opportunity can be realized by means of an element in $\mathcal{H}$ and it provides a profit for all elements of $\mathcal{S}$ (and so it is riskless).

Part of the practical relevance of the interval $[\underline{V}, \overline{V}]$ will depend of the relative sizes of the sets $(\mathcal{S}, \mathcal{H})$. On the one hand, we should design $\mathcal{S}$ to be large enough so that it allows for arbitrarily close approximations of stock charts but not any larger so as not to artificially enlarge the bounding interval. On the other hand, $\mathcal{H}$ should include only portfolios that can be implemented in practice (albeit in an idealized way) as the introduction of more powerful, but impractical, hedging strategies will artificially shrink the bounding interval.

## 2 General, Discrete, Non Probabilistic Models

**Definition 1** $\mathcal{S}(S_0)$ will be called a set of (discrete) trajectories if, for a given non-negative real number $S_0$, $\mathcal{S}(S_0)$ is a set of nonnegative real valued sequences, i.e. $S = \{S_i\}_{i \geq 0}$, $S_i \geq 0$ such that if $S, S' \in \mathcal{S}$, then $S_0 = S'_0$.

We will refer to the given $S_0$ as the *initial value* and will replace $\mathcal{S}(S_0)$ for $\mathcal{S}$ whenever convenient. Elements of such sets $\mathcal{S}$ will be simply referred as trajectories.

**Definition 2** Given a set of trajectories $\mathcal{S}(S_0)$, $\mathcal{H}$ will be called a set of portfolios acting on a set of trajectories $\mathcal{S}$ if elements $H \in \mathcal{H}$ are infinite sequences of (pairs of) functions $H = \{\Phi_i = (B_i, H_i)\}_{i \geq 0}$ with $B_i, H_i : \mathcal{S} \to \Re$. Elements $H \in \mathcal{H}$ are non anticipative, namely: $\Phi_i(S) = \Phi_i(S')$ whenever $S, S' \in \mathcal{S}$ and $S'_k = S_k$ for all $0 \leq k \leq i$. Moreover, elements $H \in \mathcal{H}$ are also required to be self-financing, namely, the following holds for all $S$ and all $i$:

$$H_i(S)S_{i+1} + B_i(S) = H_{i+1}(S)S_{i+1} + B_{i+1}(S). \tag{1}$$

Moreover, we further assume that for each $S \in \mathcal{H}$ there exists an integer $N(S)$ such that $H_k(S) = 0$ for all $k \geq N(S)$.

Elements of such sets $\mathcal{H}$ will be simply referred as portfolios.

**Definition 3** (Discrete Market Models) A discrete market model is a tuple $\mathcal{M} = (\mathcal{S}(S_0), \mathcal{H})$ where $\mathcal{S}(S_0)$ is a set of trajectories and $\mathcal{H}$ a set of portfolios.

For simplicity we will assume that $B_i$ represents the values of a Bank account with interests rates $r = 0$. Consider an unfolding stock chart $S(t)$ and bank account $B(t)$, the number $H_i(S)$ is interpreted as the hedging investment *just after* the $i$-th trading has taken place. $S_i$ is the value taken by the unfolding chart at the $i$-th trading. $B_i(S)$ is interpreted as the money in the balancing bank account *just after* the $i$-th trading has taken place. The value $S_0$ represents, for any trajectory $S$, the stock value $S(0)$ at initial time $t_0$. Given an option $Z$, i.e. a function $Z : \mathcal{S} \to \Re$, $N(S)$ represents the, path dependent, number of trading instances $i = 0, 1, \dots, N(S) - 1$ taking place until the option is exercised at instance $N(S)$ and the investment in $S$ is liquidated.

Given $S \in \mathcal{S}$ and $H \in \mathcal{H}$, the self-financing property (1) implies that the portfolio value, defined by $V_H(i, S) = B_i(S) + H_i(S)S_i$ equals:

$$V_H(i, S) = V_H(0, S_0) + \sum_{k=0}^{i-1} H_i(S)(S_{k+1} - S_k), \tag{2}$$

during the period $[i, i+1)$ for $i = 0, \dots, N(S) - 2$ and valid over $[N(S) - 1, N(S)]$ for the case $i = N(S) - 1$. Of course, $V_H(0, S_0) \equiv V_H(0, S) = B_0(S) + H_0(S)S_0$.

*Remark 1* Clearly, to specify $H = \{(B_i, H_i)\}$ one can, alternatively, specify $\{H_i\}$ and a real number $V_H(0, S)$.

The above non probabilistic notions can be extended to continuous time under the assumption that the limits of the finite sums in (2) exist as is the case of the Ito-Föllmer integral ([5]). This has been introduced and studied in [2]. The connection between the continuous time trajectories $S(t)$ and the setup of our paper is given through the use of non probabilistic stopping times $\tau_i : \mathcal{S} \to \Re$ which gives $S_i = S(\tau_i(S))$. This last remark also allows to explicitly introduce time in our present formalism (something needed for the $r > 0$ case). We refer to [1] for details.

## 2.1 Arbitrage-Free Markets and Pricing Bounds

Having the notion of a discrete financial market model we can naturally define the notion of arbitrage.

**Definition 4** Given a discrete market model $\mathcal{M} = (\mathcal{S}, \mathcal{H})$, we will call $H \in \mathcal{H}$ an arbitrage strategy if:

- $\forall S \in \mathcal{S}$, $V_H(N(S), S) \geq V_H(0, S_0)$;
- $\exists S^* \in \mathcal{S}$ satisfying $V_H(N(S^*, S^*)) > V_H(0, S_0)$.

We will say $\mathcal{M}$ is arbitrage-free if $\mathcal{H}$ contains no arbitrage strategies.

**Definition 5** A given trajectory space $\mathcal{S}$ is said to satisfy the up-down property if for $S \in \mathcal{S}$ and a positive integer $j$:

$$\sup_{\hat{S} \in \mathcal{S},\ \Pi_j(\hat{S}) = \Pi_j(S)} (\hat{S}_{j+1} - S_j) > 0, \quad \text{and} \quad \inf_{\hat{S} \in \mathcal{S}\ \Pi_j(\hat{S}) = \Pi_j(S)} (\hat{S}_{j+1} - S_j) < 0, \tag{3}$$

or:

$$\sup_{\hat{S}\in\mathcal{S},\ \Pi_j(\hat{S})=\Pi_j(S)} (\hat{S}_{j+1} - S_j) = \inf_{\hat{S}\in\mathcal{S}\ \Pi_j(\hat{S})=\Pi_j(S)} (\hat{S}_{j+1} - S_j) = 0. \tag{4}$$

**Theorem 1** *Any discrete market model $\mathcal{M} = (\mathcal{S}, \mathcal{H})$ where $\mathcal{S}$ satisfies the up-down property is arbitrage-free.*

**Definition 6** Given a discrete market model $\mathcal{M} = (\mathcal{S}, \mathcal{H})$ and a function $Z : \mathcal{S} \to \Re$, define the following quantities:

$$\overline{V}(S_0, Z, \mathcal{M}) = \inf_{H\in\mathcal{H}} \left\{ \sup_{S\in\mathcal{S}} \left\{ Z(S) - \sum_{i=0}^{N(S)-1} H_i(S)(S_{i+1} - S_i) \right\} \right\}, \tag{5}$$

and $\underline{V}(S_0, Z, \mathcal{M}) = -\overline{V}(S_0, -Z, \mathcal{M})$.

It is possible to prove that any market price for an option $Z$ that is outside of $[\underline{V}(S_0, Z, \mathcal{M}), \overline{V}(S_0, Z, \mathcal{M})]$ gives an arbitrage portfolio in the market $\mathcal{M}$.

As per Proposition 1 below, in an arbitrage-free market the zero option has zero price.

**Proposition 1** *Let $\mathcal{M}$ be an arbitrage-free discrete market model. Then*

$$\inf_{H\in\mathcal{H}} \left\{ \sup_{S\in\mathcal{S}} \left\{ - \sum_{i=0}^{N(S)-1} H_i(S)(S_{i+1} - S_i) \right\} \right\} = 0. \tag{6}$$

**Definition 7** (Bounded Discrete Market Models) A discrete market model $\mathcal{M} = (\mathcal{S}, \mathcal{H})$ will be called *n-bounded*, if there exists a natural number $n$ such that $N(S) \leq n$, for all $S \in \mathcal{S}$. Such a market will be denoted by $\mathcal{M}^n$.

Fixed $\tilde{H} \in \mathcal{H}$, $\tilde{S} \in \mathcal{S}$, and $0 \leq k \leq n$, set

$$\mathcal{S}_{(\tilde{S},k)} \equiv \{S \in \mathcal{S} : S_i = \tilde{S}_i, 0 \leq i \leq k\} \quad \text{and}$$
$$\mathcal{H}_{(\tilde{H},k)} \equiv \{H \in \mathcal{H} : H_i = \tilde{H}_i, 0 \leq i \leq k\}.$$

For $S \in \mathcal{S}$ we will use the notation $\Delta_i S \equiv S_{i+1} - S_i$ for $i \geq 0$.

The following inductive definition gives the basic dynamic programming formulation to compute the price bounds.

**Definition 8** Consider an $n$-bounded, discrete market model $\mathcal{M}^n = (\mathcal{S}, \mathcal{H})$. For a given derivative $Z$ defined on $\mathcal{S}$, any $S \in \mathcal{S}$, and any $H \in \mathcal{H}$, set:

$$\overline{U}_i\big(S, H, Z, \mathcal{M}^n\big) = 0, \quad \text{for } N(S) < i \leq n, \tag{7}$$

$$\overline{U}_{N(S)}\big(S, H, Z, \mathcal{M}^n\big) = Z(S). \tag{8}$$

For $1 \le i < N(S)$ define recursively,

$$\overline{U}_i\big(S, H, Z, \mathcal{M}^n\big) = \inf_{\tilde{H} \in \mathcal{H}_{(H,i-1)}} \sup_{\tilde{S} \in \mathcal{S}_{(S,i)}} \big[\overline{U}_{i+1}\big(\tilde{S}, \tilde{H}, Z, \mathcal{M}^n\big) + \tilde{H}_i(S)(\tilde{S}_{i+1} - S_i)\big], \tag{9}$$

and

$$\overline{U}_0\big(S_0, Z, \mathcal{M}^n\big) = \inf_{H \in \mathcal{H}} \sup_{S \in \mathcal{S}} \big[\overline{U}_1\big(S, H, Z, \mathcal{M}^n\big) + H_0(S)\Delta_0 S\big].$$

Under appropriate conditions it is possible to prove that $\overline{U}_0(S_0, Z, \mathcal{M}^n) = \overline{V}(S_0, Z, \mathcal{M}^n)$.

## References


1. Alvarez, A., Ferrando, S.: Trajectory based models, arbitrage and continuity. Preprint, December (2103)
2. Alvarez, A., Ferrando, S., Olivares, P.: Arbitrage and hedging in a non probabilistic framework. Math. Financ. Econ. **7**(1), 1–28 (2013). doi:10.1007/s11579-012-0074-5
3. Britten-Jones, M., Neuberger, A.: Arbitrage pricing with incomplete markets. Appl. Math. Finance **3**, 347–363 (1996)
4. Ferrando, S., Gonzalez, A., Degano, I., Rahsepar, M.: Discrete, non probabilistic market models. Arbitrage and pricing intervals. Preprint, December (2013)
5. Föllmer, H.: Calcul d'Itô sans probabilité. In: Seminaire de Probabilité XV. Lecture Notes in Math., vol. 850, pp. 143–150. Springer, Berlin (1981)
6. Rebonato, R.: The Hedger and the Fox, 2nd edn. Wiley, New York (2004)

# A Statistical Test for the Heston Model

**Gianna Figà-Talamanca**

**Abstract** We introduce a formal test to detect whether a times series of financial log-returns is consistent with the Heston stochastic volatility model as data generating process. The test is based on the auto-covariance structure of the integrated volatility, which is available in closed form for the model under investigation. The test suggested in this contribution also relies on the outcomes of a companion paper where we prove asymptotic results for the distribution of sample moments of the squared log-returns in the fully-specified Heston model.

**Keywords** Heston model · Sample auto-covariance · Asymptotic distribution

## 1 Introduction and Model Setting

One of the alternative approaches for the generalization of the seminal paper by [2] on option pricing theory, is to allow for random volatility in the underlying stock price process (see [8, 11, 12] and [7], among others). Such models, usually referred to as Stochastic Volatility models, may explain many *stylized facts* in the stock and in the derivative markets, such as the leptokurtosis of financial log-returns and the so-called smile curve of the implied volatility of options when plotted against the strike price (see e.g. [3]). Even though classical Stochastic Volatility models are outdated from a modelling viewpoint, a renewed attention has been recently devoted to the Heston model ([7]) for which a quasi-closed formula is available for the price of European *plain vanilla* options. This property make it possible to calibrate the Heston model parameters on a sample of market prices for derivative financial instruments. Calibration of parameters may be performed on market option prices as suggested, among others, in [4] but may be applicable considering more complex derivatives, such as volatility and variance options and swaps, for which closed formulas have been made available in recent years (see e.g. [10]) under Heston assumptions.

G. Figà-Talamanca (✉)
Dipartimento di Economia, University of Perugia, via Pascoli 23, Perugia, Italy
e-mail: gianna.figatalamanca@unipg.it

C. Perna, M. Sibillo (eds.), *Mathematical and Statistical Methods for Actuarial Sciences and Finance*, DOI 10.1007/978-3-319-05014-0_24, 

Motivated by this new interest and by the huge increasing of estimation and calibration techniques for the Heston model, often leading to very different parameters values, we introduce in this paper a formal statistical test to detect whether the auto-covariance structure for the squared log-returns of a given financial stock is consistent with the auto-covariance structure for the integrated volatility process which is available in closed form for the Heston model (see [6]). The Test statistics is built upon the outcomes detailed in [5] which are omitted here for brevity. In the rest of the paper we consider the Heston model, described by the following bi-dimensional process:

$$\begin{aligned} dY_t &= \sqrt{V_t} d\widetilde{W}_t, \quad Y_0 = 0, \\ dV_t &= \alpha(\beta - V_t)dt + c\sqrt{V_t} dW_t, \quad V_0 = \nu, \end{aligned} \tag{1}$$

where $(\widetilde{W}, W)$ is a Brownian motion in $\mathbb{R}^2$ defined on a probability space $(\Omega, F, P)$ with $\langle d\widetilde{W}, dW \rangle = \rho dt$ and $\nu$ is a real random variable defined on $\Omega$, independent of $(\widetilde{W}, W)$; $\beta$, $\alpha$ are constant parameters representing, respectively, the long-run mean and the mean-reversion speed (toward $\beta$) of the instantaneous variance $V_t$ and $c$ is the so-called "volatility of volatility" parameter. If $\alpha > 0$ and $2\alpha\beta \geq c^2$, then it is well known that $V_t$ is a strictly stationary Markov process on the state space $(l, r) = (0, +\infty)$.

## 2 A Formal Test for the Heston Model

Assume we are given a time series $\{Y_0, Y_1, \ldots, Y_n\}$ for the log-price process of a financial stock, with observation step $\Delta$. We want to test whether this time-series may have been generated from a continuous process as defined in (1) with assigned parameters. Define, for $i = 1, 2, \ldots, n$, the scaled log-returns $\{R_i\}_i$ and the sample moments $M_h$, for $h \geq 0$, as follows:

$$R_i = \frac{Y_i - Y_{i-1}}{\sqrt{\Delta}}, \tag{2}$$

$$M_0 = \frac{1}{n} \sum_1^n R_i^2, \tag{3}$$

$$M_h = \frac{1}{n} \sum_1^{n-h} R_i^2 R_{i+h}^2. \tag{4}$$

Let $\mathbf{M} = (M_0, M_1, M_2, \ldots, M_H)$ for a fixed lag $H$; the results in [5] guarantee that, if the discrete process $\{Y_i\}_i$ is a discretized version of the continuous process defined by (1) then, for $n$ sufficiently large, $\mathbf{M}$ is asymptotically multivariate normal with mean $\boldsymbol{\mu}$ and covariance matrix $\mathbf{n\Psi}$ (for details on how to obtain $\boldsymbol{\mu}$ and $\boldsymbol{\Psi}$ see [5]).

Set $\mathbf{A} = \boldsymbol{\Psi}^{-1}$; by applying Theorem 5.4.2 in [9] we get that the asymptotic distribution of

$$\sqrt{n}\mathbf{A}^{1/2}(\mathbf{M} - \boldsymbol{\mu}) \tag{5}$$

is a standard multivariate normal of dimension $H + 1$. Further, the quadratic form

$$Q_H^{(n)} = n(\mathbf{M} - \boldsymbol{\mu})^T \mathbf{A}(\mathbf{M} - \boldsymbol{\mu}) \tag{6}$$

is distributed as a $\chi^2$ with $H + 1$ degrees of freedom.

Hence, we can use the $Q_H^{(n)}$ statistics in order to construct a formal statistical test for the sample auto-covariance structure of the process $\{R_i^2\}_i$ under the Null of a data generating process for $\{Y_i\}_i$ defined by (1). Precisely, if $F_{\chi^2(H+1)}$ represents the cumulative distribution function of a $\chi^2$ random variable with $H + 1$ degrees of freedom and $n$ is sufficiently large, we reject the Null hypothesis with a confidence level $p$ when $Q_H^{(n)} > F_{\chi^2(H+1)}^{-1}(p)$.

While $\boldsymbol{\mu}$ is easily available in explicit form, the derivation on matrix $\boldsymbol{\Psi}$ in closed form (with respect to the assigned parameters) is often cumbersome, although possible in principle. Nevertheless, it might be replaced by an estimate of the matrix obtained, for example, via the Bootstrap resampling method; if the estimator is consistent then the asymptotic results on the distribution of $Q_H^{(n)}$ still hold (see [9]).

## 3 A Simulation Exercise

As a preliminary analysis, in order to evaluate the performance of the proposed test, we simulate $m = 1000$ paths for a one year long time series of log-prices with discretization step $\delta = 1/3000$, assuming $\alpha = 1$, $\beta = 0.05$, $c = 0.25$ and $\rho = -0.04$; for the simulation we applied the Quadratic-Exponential scheme proposed in [1] with $Y_0 = \log 100$ and $\nu = 0.05$.

For each $j = 1, 2, \ldots, m$, we compute the scaled log-returns $\{R_i^j\}_i$ and the sample empirical moments to get $\mathbf{M}^j$; $\boldsymbol{\mu}$ is obtained via the explicit formulas with respect to model parameters (see [5]) while, for the sake of simplicity, the matrix $\mathbf{n}\boldsymbol{\Psi}$ is derived as the covariance matrix of $\mathbf{M}^j$ on all simulated paths. Finally, we evaluate the statistics $Q_{H,j}^{(n)}$, with $H = 5$, and apply the test suggested above to each simulated trajectory; as an example we represent in Fig. 1 the sample moments in $\mathbf{M}^j$ for two sample trajectories as well as their theoretical limit $\boldsymbol{\mu}$ and the corresponding asymptotic confidence bands with 95 % confidence level.

As it was expected, the test do not reject the Heston model specification neither for $p = 95$ % nor for $p = 99$ %. Analogous results are obtained if we replace matrix $\mathbf{n}\boldsymbol{\Psi}$ with the Covariance matrix obtained through a Bootstrap resampling of each path. Of course, theoretical properties of the test should be further investigated and more extensive simulation experiments be performed. We leave these issues to future research. Anyway, the preliminary outcomes reported here show that the pro-

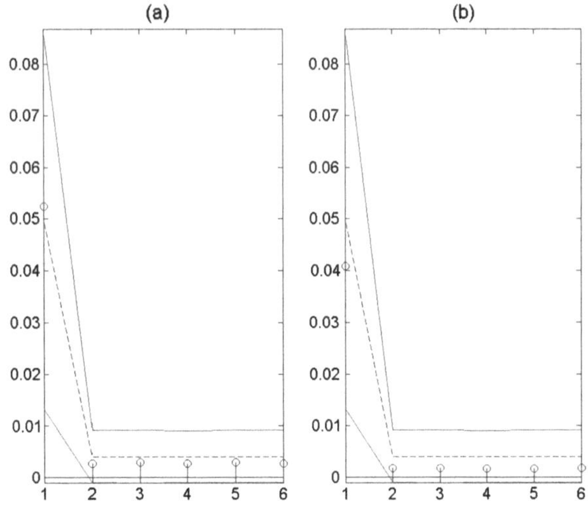


**Fig. 1** Sample auto-covariance structure (circles) for two of the simulated paths for $\alpha = 1$, $\beta = 0.05$, $c = 0.25$, $\rho = -0.04$, $Y_0 = \log 100$ and $\nu = 0.05$. The mean value $\mu_M$ (*dashed-line*) and confidence bands (*solid line*) are also reported

posed test is a promising tool in order to test the Heston model on market historical data; this might be useful, as already observed in the introduction, when models parameters are obtained implicitly using derivatives prices as input data instead of historical prices of the underlying to which parameters are indeed related. We are currently working on the application of the test to observed financial stock prices and the results will contribute to a related paper.

## References

1. Andersen, L.B.G.: Simple and efficient simulation of the Heston stochastic volatility model. J. Comput. Finance **11**(3), 1–42 (2008)
2. Black, F., Scholes, M.: The pricing of options and corporate liabilities. J. Polit. Econ. **81**, 637–659 (1973)
3. Cont, R.: Empirical properties of asset returns: stylized facts and statistical issues. Quant. Finance **1**(2), 223–236 (2001)
4. Ewald, C.O., Zhang, A.: A new method for the calibration of stochastic volatility models: the Malliavin gradient method. Quant. Finance **6**(2), 147–158 (2006)
5. Figà-Talamanca, G.: Limit results for stochastic volatility models. Quaderni del Dipartimento di Economia, Finanza e Statistica 63 (2008)
6. Figà-Talamanca, G.: Testing volatility autocorrealtion in the CEV model. Comput. Stat. Data Anal. **53**, 2201–2218 (2009)
7. Heston, S.L.: A closed-form solution for options with stochastic volatility with applications to bond and currency options. Rev. Financ. Stud. **6**, 327–343 (1993)
8. Hull, J., White, A.: The pricing of options on assets with stochastic volatility. J. Finance **42**(2), 281–300 (1987)
9. Lehmann, E.L.: Elements of Large-Sample Theory. Springer, Berlin (1998)
10. Sepp, A.: Pricing options on realized variance in the Heston model with jumps in returns and volatility. J. Comput. Finance **11**(4), 33–70 (2008)
11. Stein, E., Stein, J.: Stock price distributions with stochastic volatility. Rev. Financ. Stud. **4**, 27–752 (1991)
12. Wiggins, J.: Option values under stochastic volatilities. J. Financ. Econ. **19**(351), 372 (1987)

# Threshold Random Walk Structures in Finance

**Francesco Giordano, Marcella Niglio, and Cosimo Damiano Vitale**

**Abstract** In this paper we propose a new model that generalizes, in nonlinear domain, the random walk process: we call this model *threshold random walk*. From the empirical point of view it is able to model the asymmetric behaviour of financial data that is neglected from the random walk structure. We further provide a statistical tool for testing unit root versus a stationary threshold autoregressive model.

**Keywords** Threshold random walk · Unit root test

## 1 Introduction

The study of stochastic processes that are able to mimic the dynamic structure of stock returns has largely interested the statistical and econometric literature. In this context the well known Random Walk has been historically used by many modelers that have based their theoretical assumptions on the market efficiency and so on the unpredictability of stock returns on the basis of current available information.

The complexity of the stock markets and the relationships among their agents have made quite weak the efficiency assumption and a large number of stochastic structures have been proposed to model the behaviour of most financial data.

The aim of the present paper is to propose a generalization, in nonlinear domain, of the Random Walk model. More precisely starting from the threshold models, widely discussed in [14], we present the *Threshold Random Walk* model. In more detail in the next section, after shortly review the more recent literature on the threshold models with unit root, we propose a new stochastic model that is able to take into account the asymmetric behaviour of financial data. A test for detecting the presence of a unit root against model stationarity is further provided.

F. Giordano · M. Niglio (✉) · C.D. Vitale
Di.S.E.S., University of Salerno, via Giovanni Paolo II 132, 84084 Fisciano (SA), Italy
e-mail: mniglio@unisa.it

F. Giordano
e-mail: giordano@unisa.it

C.D. Vitale
e-mail: cvitale@unisa.it

C. Perna, M. Sibillo (eds.), *Mathematical and Statistical Methods for Actuarial Sciences and Finance*, DOI 10.1007/978-3-319-05014-0_25,

## 2 Unit Roots in Threshold Autoregressive Models

Let $X_t$ be a nonlinear time series, it is said to be generated by a threshold autoregressive (TAR) process when:

$$X_t = \sum_{j=1}^{k} \left( \sum_{i=1}^{p} \phi_i^{(j)} X_{t-i} \right) I(Y_{t-d} \in \mathcal{R}_j) + e_t, \qquad (1)$$

where $k$ is the number of regimes, $p$ is the autoregressive order, $I(\cdot)$ is an indicator function, $d$ is the threshold delay, $\mathcal{R}_j$ is a subset of $\mathcal{R}$, such that $\mathcal{R}_j = [r_{j-1}, r_j)$ and $-\infty = r_1 < r_2 < \cdots < r_{k-1} < r_k = +\infty$, and $e_t$ is the error term. If $Y_{t-d} = X_{t-d}$ model (1) is said to be a Self Exciting Threshold AutoRegressive (SETAR) model.

The study of unit roots in presence of SETAR processes has been differently faced: after the results in [12] where the loss of power of the Dickey and Fuller ([4]) test is shown, Enders and Granger [5] proposes a unit root testing procedure for a SETAR process with $k = 2$, $p = 1$ and $d = 1$.

A seminal contribution in unit root testing in presence of threshold processes has been given in [3] for the threshold autoregressive model (1) with $k = 2$ and $d = 1$ whereas [7] propose the so called threshold unit root model, given by (1) with $p = 1$. Bec et al. [1], after discussing the stationarity conditions of model (1) with $Y_{t-d} = X_{t-d}$ and $k = 3$, proposes a unit root test for a particular SETAR model then generalized in [2]. Further results are given in [8] that proposes a unit root test for the following three regimes SETAR structure:

$$X_t - X_{t-1} = \rho_1 X_{t-1} I_{\{X_{t-1} \le r_1\}} + X_{t-1} I_{\{r_1 < X_{t-1} \le r_2\}} + \rho_3 X_{t-1} I_{\{X_{t-1} > r_2\}} + e_t, \quad (2)$$

where the second regime is assumed to have a unit root. Kapetanios and Shin [8] generalizes the testing results even to the case where $e_t$ is a linear stationary autoregressive process, $e_t = \sum_{i=1}^{p} \gamma_i \nabla X_{t-i} + \varepsilon_t$ (with $\varepsilon_t \sim$ i.i.d.$(0, \sigma^2)$).

Another contribution in this domain is given in [13] that starting from the following SETAR model:

$$\nabla X_t = \rho_1 X_{t-1} I_{\{X_{t-1} \le r_1\}} + \rho_2 X_{t-1} I_{\{X_{t-1} > r_1\}} + e_t, \qquad (3)$$

proposes an Augmented Dickey and Fuller-type unit root test where $e_t$ is allowed to be weakly dependent.

Further results on unit root in presence of three regimes threshold autoregressive models are given in [10] whereas new results mainly related to the estimation of the model parameters are given in [9] and [6].

Starting from these results we introduce the following model: let $\nabla X_t = X_t - X_{t-1}$ be a threshold process with two regimes and autoregressive order $p = 1$:

$$\nabla X_t = \rho_1 X_{t-1} I_{t-1} + \rho_2 X_{t-1} (1 - I_{t-1}) + e_t, \qquad (4)$$

with $\rho_j = \phi_1^{(j)} - 1$ (for $j = 1, 2$), $I_{t-1} = 1$ if $X_{t-1} \ge r$ and 0 otherwise, where the threshold value $r$ belongs to the compact set $P \subset \mathcal{R}$ and the innovation process $\{e_t\}$ can follow two alternative threshold structures. The first one is

$$e_t = e_{1t} I'_{t-1} + e_{2t} (1 - I'_{t-1}), \qquad (5)$$

with indicator function $I'_{t-1} = 1$ if $\nabla X_{t-1} \geq r'$ and 0 otherwise, whereas $\{e_{1t}\}$ and $\{e_{2t}\}$ are two independent *i.i.d.* sequences with positive density functions for both $e_{1t}$ and $e_{2t}$ with $E(e_{1t}) = E(e_{2t}) = 0$ and $Var(e_{1t}) = \sigma_1^2 < \infty$, $Var(e_{2t}) = \sigma_2^2 < \infty$.

The second structure for the innovation process is

$$e_t = \beta_1 \nabla X_{t-1} I'_{t-1} + \beta_2 \nabla X_{t-1} (1 - I'_{t-1}) + \varepsilon_t \tag{6}$$

with $\beta_1 < 1$, $\beta_2 < 1$ and $\beta_1 \beta_2 < 1$ as in [11].

When $\rho_1 = \rho_2 = 0$ we call model (4) *Self Exciting Threshold Random Walk* (SETRW).

Starting from this theoretical context we propose a unit root testing procedure for model (4) with innovations (5) or (6), which is based on a Wald statistic. In more detail given model (4) with innovations (5), it can be equivalently written as:

$$\nabla \mathbf{X} = \mathbf{Y} \boldsymbol{\rho} + \mathbf{e}, \tag{7}$$

where $\boldsymbol{\rho} = (\rho_1, \rho_2)^T$, $\nabla \mathbf{X} = (\nabla X_1, \ldots, \nabla X_n)^T$, $\mathbf{e} = (e_1, \ldots, e_n)^T$, $n$ is the time series length and

$$\mathbf{Y} = \begin{pmatrix} X_0 I_0 & , & X_0(1 - I_0) \\ \vdots & , & \vdots \\ X_{n-1} I_{n-1} & , & X_{n-1}(1 - I_{n-1}) \end{pmatrix}.$$

The Null hypothesis that $\rho_1 = \rho_2 = 0$ can be tested using the Wald statistic related to Eq. (7)

$$\mathcal{W}(r, r') = \frac{\hat{\boldsymbol{\rho}}^T (\mathbf{Y}^T \mathbf{Y}) \hat{\boldsymbol{\rho}}}{S_e^2}, \tag{8}$$

where $\hat{\boldsymbol{\rho}}$ is the OLS estimator of $\boldsymbol{\rho}$ and $S_e^2 = \frac{1}{n-2} \sum_{t=1}^{n} \hat{e}_t^2$ with $\hat{e}_t$ the estimated residuals from model (4) with innovation process in (5).

To obtain the asymptotic distribution of $\mathcal{W}(r, r')$ consider the following assumption:

**Assumption 1** *Let* $\{e_{1t}\}$ *and* $\{e_{2t}\}$ *be two independent i.i.d. sequences of random variables with zero mean,* $0 < \sigma_1^2$, $0 < \sigma_2^2$ *variances and* $E(|e_{1t}|^{4+\eta}) < \infty$, $E(|e_{2t}|^{4+\eta}) < \infty$, *respectively for some* $\eta > 0$. *Moreover, the density functions are positive on* $\mathcal{R}$.

In presence of fixed thresholds $r$ and $r'$ we can state:

**Proposition 1** *Given the threshold process* (4) *that fulfills the assumption* 1, *under the Null hypothesis* $\rho_1 = \rho_2 = 0$ *the Wald statistic defined by* (8) *with* $r = r' = 0$, *has the following asymptotic distribution*

$$\mathcal{W}(0,0) \xrightarrow{d} \frac{[\int_0^1 I_{\{B(s) \geq 0\}} B(s) dB(s)]^2}{\int_0^1 I_{\{B(s) \geq 0\}} B^2(s) ds} + \frac{[\int_0^1 I_{\{B(s) < 0\}} B(s) dB(s)]^2}{\int_0^1 I_{\{B(s) < 0\}} B^2(s) ds},$$

*where* $B(s)$ *is the standard Brownian motion with* $s \in [0, 1]$.

Similar results can be obtained for model (4) with innovations (6) that for brevity are here omitted.

## References

1. Bec, F., Salem, M.B., Carrasco, M.: Tests for unit-root versus threshold specification with an application to the purchasing power parity relationship. J. Bus. Econ. Stat. **22**, 382–395 (2004)
2. Bec, F., Guay, A., Guerre, E.: Adaptive consistent unit-root tests based on autoregressive threshold model. J. Econom. **142**, 94–133 (2008)
3. Caner, M., Hansen, B.: Threshold autoregression with a unit root. Econometrica **69**, 1555–1596 (2001)
4. Dickey, D.A., Fuller, W.A.: Distribution of the estimators for autoregressive time series with a unit root. J. Am. Stat. Assoc. **74**, 427–431 (1979)
5. Enders, W., Granger, C.W.J.: Unit-root tests and asymmetric adjustment with an example using the term structure of interest rates. J. Bus. Econ. Stat. **16**, 304–311 (1998)
6. Gao, J., Tjøstheim, D., Yin, J.: Estimation in threshold autoregressive models with a stationary and a unit root regime. J. Econom. **172**, 1–13 (2013)
7. Gonzalez, M., Gonzalo, J.: Threshold unit root models. Working Paper, University Carlos III de Madrid (1997)
8. Kapetanios, G., Shin, Y.: Unit root tests in three-regime SETAR model. Econom. J. **9**, 252–278 (2006)
9. Liu, W., Ling, S., Shao, Q.M.: On non-stationary threshold autoregressive models. Bernoulli **17**, 969–986 (2011)
10. Maki, D.: Tests for a unit root using three-regime TAR models: power comparison and some applications. Econom. Rev. **28**, 335–363 (2009)
11. Petruccelli, J.D., Woolford, S.W.: A threshold AR(1) model. J. Appl. Probab. **21**, 270–286 (1984)
12. Pippenger, M.K., Gregory, E.: A note on the empirical power of unit root tests under threshold processes. Oxf. Bull. Econ. Stat. **55**, 473–481 (1993)
13. Seo, M.H.: Unit root test in a threshold autoregression: asymptotic theory and residual-based block bootstrap. Econom. Theory **24**, 1699–1716 (2008)
14. Tong, H.: Non-Linear Time Series. A Dynamical System Approach. Clarendon Press, Oxford (1990)

# Stochastic Mortality Models. Application to CR Mortality Data

**Ján Gogola**

**Abstract** The ageing process is a great challenge for many European countries, not excluding Czech Republic (CR) and it brings financial risk in areas such as social policy, pensions and health care. The motivation for this paper is to compare various mortality models. We have attempted to explain mortality improvements for males aged 62–90 in CR using a several stochastic mortality models. We compare quantitatively number of stochastic models explaining improvements in mortality rates in CR. It is clear that mortality improvements are driven by an underlying process that is stochastic. Numbers of stochastic models have been developed to analyse these mortality improvements. We will deal in models such as Lee-Carter model, Renshaw and Haberman model, Aged-Periodic-Cohort model (APC), Cairns-Blake-Dowd model (CBD) and their extensions. Each model is fitted to the male data between 1968 and 2011. Our analysis focuses on mortality at higher ages (62–90), given our interest in pension-related applications. By the Bayes Information Criterion (*BIC*) we find that an extension of the Cairns-Blake-Dowd (CBD) model fits the Czech Republic male's data best.

**Keywords** Mortality · Constraints · Bayes Information Criterion · Force of mortality

## 1 Introduction

This paper deals with a stochastic mortality models. In many developed countries the life expectancy is increasing. Such increases imply rapidly increasing cost in pensions and health care for the elderly. The effort of forecasting the future trend of mortality has been a subject of great interest. Over the past twenty years, a great number of new approaches were developed in order to forecast mortality by using stochastic models, such as the ones presented by McNown and Rogers (1989, 1992), Bell and Monsell (1991), and Lee and Carter (1992).

J. Gogola (✉)
Institute of Mathematics and Quantitative methods, Faculty of Economics and Administration, University of Pardubice, Studentská 84, 532 10 Pardubice, Czech Rebublic
e-mail: jan.gogola@upce.cz

C. Perna, M. Sibillo (eds.), *Mathematical and Statistical Methods for Actuarial Sciences and Finance*, DOI 10.1007/978-3-319-05014-0_26,

The Lee-Carter model became one of the most well-known models and it is applied in different countries around the world to forecast age-specific death rates. In their paper [4] they used mortality data classified by age of death and year of death, and then modelled the force of mortality.

A wide family of mortality models is assessed by [1]. They consider eight models for mortality rates in England and Wales. They have noticed if these models are applied to different countries then conclusions about which model is most suitable might be different.

The primary aim of this paper is to apply some of these stochastic models on age-specific death rates from CR and compare and rank these varieties of models. We are focusing on mortality at higher ages (62–90), given our interest in pension-related application. This paper follows on the article [3]. They deal with the development and the prediction of life expectancy in selected European countries (CR, Slovakia, Finland, Spain) by applying only Lee-Carter model.

Our approach consists of three steps: (1) obtain and solve the maximum likelihood equations for each model; (2) write the associated code in **R** program; (3) select the best fitting model. The plan of this paper is as follow. In Sect. 2 is a brief description of our data and notation. In Sect. 3 we describe six stochastic mortality models. Section 4 gives the main results. We are fitting models to our data. Finally, in Sect. 5 we present some conclusions.

## 2 Data and Notation

We use data on male deaths and exposure to risk between 1968 and 2011 from the Human Mortality Database. We consider the restricted age range from 62 to 90, the range of interest to providers of pensions.

The data will cover the range $x_1, x_2, \ldots, x_{n_a}$ and $t_1, t_2, \ldots, t_{n_y}$, with unit increments where $n_a$ is the number of ages and $n_y$ is the number of years. Let $d_{x,t}$ be the number of deaths aged $x$ last birthday in calendar year $t$. We suppose that the data on deaths are arranged in a matrix $\mathbf{D} = (d_{x,t})$. Similarly, the data on exposure to risk are arranged in a matrix $\mathbf{E}^c = (e_{x,t})$ where $e_{x,t}$ is the average size of the population aged $x$ last birthday in calendar year $t$. We denote the force of mortality at exact time $t$ for lives with exact age $x$ by $\mu_{x,t}$. The force of mortality is interpreted as the instantaneous death rate and the probab. that individual dies in the small interval $(t, t + \mathrm{d}t)$ is approx. $\mu_{x,t} \cdot \mathrm{d}t$. We also consider the mortality rate $q_{x,t}$. This is the probab. that an individual aged exactly $x$ at exact time $t$ will die between $t$ and $t + 1$.

Some of the models we consider model the force of mortality $\mu_{x,t}$, whereas others model the mortality rate $q_{x,t}$. To ensure a valid comparison between the different models, our analysis of the models for $q_{x,t}$ involve an additional step. For a given set of parameters we calculate the $q_{x,t}$ then we transform these into force of mortality using the identity $\mu_{x,t} = \log(1 - q_{x,t})$.

We can calculate the likelihood for all models consistently based on the $\mu_{x,t}$.

For a given model we use $\phi$ to represent the full set of parameters and the notation for $\mu_{x,t}$ is extended to $\mu_{x,t}(\phi)$, to indicate its dependence on these parameters.

For all models the log-likelihood is

$$l(\phi; D, E^c) = \sum_{x,t} \left( d_{x,t} \cdot \log\left[ e_{x,t} \cdot \mu_{x,t}(\phi) \right] - e_{x,t} \cdot \mu_{x,t}(\phi) - \log(d_{x,t}!) \right), \quad (1)$$

and estimation is by maximum likelihood.

## 3 The Mortality Models

We deal with the following models:

1. Lee-Carter model;
2. APC model [2];
3. Renshaw-Haberman model [5];
4. Cairns-Blake-Dowd model (CBD model);
5. CBD model with cohort effect;
6. Quadratic CBD model with cohort effect.

Our models are fitting to historical data.

With nested models we have mentioned above we can always improve the maximum likelihood by introducing further parameters into our model.

We penalize the likelihood each time we add in additional parameters.

One of the well-known approach is the Akaike Information Criterion (*AIC*) which seeks to maximize $l_i(\hat{\phi}_i; x) - k_i$.

With the *AIC* we can see that the addition of one parameter requires a improvement in the log-likelihood of at least 1 for the more complex model to be worthwhile.

We are focusing on the Bayes Information Criterion (*BIC*).

## 4 Main Results

The maximum log-likelihood of our models are displayed in Table 1, together with the respective values of *BIC*.

If we rank models with the top model having the maximum log-likelihood, we can see that Quadratic CBD model with cohort effects is the best. However if we rank models with the top model having the lowest *BIC*, it can be seen that the order has changed and CBD model with cohort effects comes out on top.

## 5 Conclusion

We have attempted to explain mortality improvements for males aged 62–90 in CR using a number of stochastic mortality models. We have found out that different models have different strengths. By the value of maximum log-likelihood the Quadratic CBD model with cohort effects fits our data set best.

By the *BIC* ranking criteria, the CBD model with cohort effects fits our data set best.

**Table 1** Models with their maximum log-likelihood, effective number of parameters e. n. p., and Bayes Information Criteria

| Model | Maximum log-likelihood | Rank | e. n. p. | *BIC* | Rank |
|---|---|---|---|---|---|
| Lee-Carter | −7346.371 | 5. | 100 | 15407.890 | 5. |
| Renshaw-Haberman | −6408.408 | 2. | 199 | 14239.960 | 3. |
| APC | −6922.780 | 4. | 142 | 14861.070 | 4. |
| Cairns-Blake-Dowd | −8082.376 | 6. | 88 | 16794.080 | 6. |
| CBD with cohort effects | −6414.079 | 3. | 158 | **13978.090** | **1.** |
| Quadratic CBD | **−6351.361** | **1.** | 201 | 14140.170 | 2. |

Source: Own Processing

# References


1. Cairns, A.J.G., Blake, D., Dowd, K., Coughlan, G.D., Epstein, D., Ong, A., Balevich, I.: A quantitative comparison of stochastic mortality models using data from England and Wales and the United States. N. Am. Actuar. J. **13**(1), 1–35 (2009)
2. Currie, I.D., Durban, M., Eilers, P.H.C.: Smoothing and forecasting mortality rates. Stat. Model. **4**, 279–298 (2004)
3. Jindrová, P., Slavíček, O.: Life expectancy development and prediction for selected European countries. In: 6-th International Scientific Conference Managing and Modelling of Financial Risk Proceedings, pp. 303–312. VŠB-TU, Ostrava (2012). ISBN 978-80-248-2835-0
4. Lee, R.D., Carter, L.: Modelling and forecasting the time series of U.S. mortality. J. Am. Stat. Assoc. **87**, 659–671 (1992)
5. Renshaw, A.E., Haberman, S.: A cohort-based extension to Lee-Carter model for mortality reduction factors. Insur. Math. Econ. **38**, 556–570 (2006)

# Risk Adjusted Dynamic Hedging Strategies

**Martin Harcek**

**Abstract** The aim of the paper is to develop a dynamic portfolio hedging strategy leading to an optimal wealth policy in a finite investment horizon while obeying a risk constraint. The utility maximization problem is restricted by an upper bound applied on the Conditional Value-at-Risk (CVaR) measure. We investigate the strategy dynamics and properties in terms of the desired wealth distribution and risky assets exposure.

**Keywords** Dynamic strategy · Conditional Value-at-Risk · Complete market

## 1 Market Settings

We consider a financial market with $N$ *risky* assets with random returns and one risk-free asset with deterministic yield. The dynamics of market prices follow the system of $N$ stochastic and one ordinary differential equations $dS(t) = S(t)\mu(t)dt + S(t)\sigma(t)dw(t)$ and $dB(t) = B(t)r(t)dt$, where $\mu(t)$ is the vector of drifts, $\sigma(t)$ is the volatility matrix and $r(t)$ is the deterministic bond yield. The process $w(t)$ is an $N$-dimensional standardised Brownian motion.

Let $W(t)$ be a value of the portfolio at time $t$. The investor chooses the investment horizon $T$ and the investment strategy $\theta(t)$, which represents the fraction of wealth invested in each risky asset at time $t \in (0, T]$. Then the portfolio value follows the stochastic differential equation $dW(t) = W(t)\theta(t)^\top(\mu(t)dt + \sigma dw(t)) + W(t)(1 - \theta(t)^\top \underline{1})r(t)dt$, where $\underline{1} \equiv (1, 1, \dots, 1)^\top$. We assume that market is *complete*. Such an assumption implies (by Ito's lemma) the existence of a unique state-price density process $\xi(t)$ given by $d\xi(t) = -\xi(t)r(t)dt - \xi(t)\kappa(t)^\top dw(t)$, where $\xi(0) = 1$ and $\kappa(t) = \sigma(t)^{-1}(\mu(t) - r(t)\underline{1})$ is the *Sharpe ratio* process.

---

M. Harcek (✉)
Dep. of Applied Mathematics and Statistics, Faculty of Mathematics, Physics and Informatics, Comenius University, Mlynska Dolina, 842 48 Bratislava, Slovakia
e-mail: martin@harcek.com

C. Perna, M. Sibillo (eds.), *Mathematical and Statistical Methods for Actuarial Sciences and Finance*, DOI 10.1007/978-3-319-05014-0_27, 

## 2 Problem Statement

The coherent [1] and convex [3] risk measure *Conditional Value-at-Risk* (CVaR) is defined as a conditional expectation of losses greater than the Value-at-Risk (VaR) threshold. VaR is a widely used risk measure, technically it is equal to $(1-\alpha)$-quantile of the portfolio loss distribution (e.g. 99 %)

$$
\begin{aligned}
CVaR_\alpha(W_0 - W_T) &= \mathbb{E}\big[W_0 - W_T \,|\, W_0 - W_T \geq VaR_\alpha(W_0 - W_T)\big] \leq \delta W_0 \\
VaR_\alpha(W_0 - W_T) &= \big\{c \in \mathbb{R} : \mathbb{P}(W_0 - W_T \leq c) = 1 - \alpha\big\}.
\end{aligned}
$$

As the above statement is relatively complex, we substitute it by a more convenient representation

$$
G_\alpha(W_0 - W_T, c) = c + \frac{1}{\alpha}\int_{-\infty}^{\infty} (W_0 - W_T - c)^+ dP(W_0 - W_T). \tag{1}
$$

The way of CVaR substitution is well described in [4]. We incorporate the risk constraint in terms of the terminal portfolio CVaR in the utility maximization problem, hence define the *CVaR Investor Optimization Problem*:

$$
\max_{W_T, c} \mathbb{E}\big[u(W_T)\big] \quad \text{s.t.} \quad \mathbb{E}[\xi_T W_T] \leq W_0 \quad \text{and} \quad G_\alpha(W_0 - W_T, c) \leq \delta W_0, \tag{2}
$$

where $G_\alpha(W_0 - W_T, c)$ is given by (1), $u(\cdot)$ is the utility function, $W_0$ is the initial wealth, $\alpha$ and $\delta$ are given exogenously, $\xi_T$ is defined in previous section and $c \in \mathbb{R}$ is a variable to be optimized.

## 3 Optimal Investment Strategy

We solve the problem as a two-stage optimization procedure. The solution of the first stage (Theorem 1) defines an optimal portfolio choice in the $W_T \times \xi_T$ space for each given $c$. As a result of the second stage optimization, we obtain an optimum through all possible settings of $c$ by solving $\max_{c \in \mathbb{R}} \mathbb{E}[u(\hat{W}_T(c))]$. In our practical calculations we suppose that the investor's preferences are well described by an iso-elastic utility function given by $u(x) = \frac{x^p}{p}$, $p < 0$ and the exogenous model parameters $r$ and $\kappa$ are constant in time.

**Theorem 1** ($T$-Time Optimal Portfolio Choice) *Define*

$$
W_T(c, y_1, y_2) = I(y_1\xi_T)1_{\{\xi_T < \underline{\xi}\}} + (W_0 - c)1_{\{\underline{\xi} \leq \xi_T < \overline{\xi}\}} + I(y_1\xi_T - y_2/\alpha)1_{\{\overline{\xi} \leq \xi_T\}},
$$

*where* $c \in \mathbb{R}$, $y_1 > 0$, $y_2 \geq 0$, $I(\cdot)$ *is the inverse function of* $u'(\cdot)$, $1_{\{\cdot\}}$ *is the indicator function,* $\underline{\xi} = u'(W_0 - c)/y_1$ *and* $\overline{\xi} = (u'(W_0 - c) + \frac{y_2}{\alpha})/y_1$. *Denote* $\hat{y}_1$ *and* $\hat{y}_2$ *to be a solution of equation system*

$$
\begin{aligned}
\mathbb{E}\big[\xi_T W_T(c, \hat{y}_1, \hat{y}_2)\big] &= W(0) \\
c + \frac{1}{\alpha}\mathbb{E}\big[\big(W_0 - W_T(c, \hat{y}_1, \hat{y}_2) - c\big)^+\big] &= \delta W_0 \quad \text{or} \quad \hat{y}_2 = 0.
\end{aligned}
$$

*Then* $\forall c$ *the problem* (2) *attains maximum at the point* $\hat{W}_T(c) \equiv W_T(c, \hat{y}_1, \hat{y}_2)$.

**Theorem 2** ($t$-Time Optimal Portfolio Choice) *The wealth process of the solution* $\hat{W}_T$ *given by Theorem* 1 *is*

$$W(t) = \frac{y_1^{\frac{1}{p-1}}}{\xi_t} e^{\frac{p}{p-1}(\ln \xi_t + (\frac{\|\kappa\|^2}{2p-2} - r)(T-t))} \Phi(d_3)$$
$$+ \frac{W_0 - c}{\xi_t} e^{\ln \xi_t - r(T-t)} \big(\Phi(d_2) - \Phi(d_1)\big)$$
$$+ \frac{1}{\xi_t} \int_{\overline{\xi}}^{\infty} \xi_T \left( y_1 \xi_T - \frac{y_2}{\alpha} \right)^{\frac{1}{p-1}} d\mathbb{P}(\xi_T),$$

*where* $\Phi(\cdot)$ *is the cumulative distribution function of* $N(0,1)$,

$$d_1 = \frac{\ln \underline{\xi} - \ln \xi_t + (r - \frac{1}{2}\|\kappa\|^2)(T-t)}{\|\kappa\| \sqrt{T-t}}, \qquad d_2 = d_1 + \frac{\ln \overline{\xi} - \ln \underline{\xi}}{\|\kappa\| \sqrt{T-t}} \quad \textit{and}$$

$$d_3 = d_1 - \frac{p}{p-1} \frac{\|\kappa\|}{2} \sqrt{T-t}.$$

By definition of the portfolio wealth process $W_t$ and the optimal process of the solution $\hat{W}_T$ given by Theorem 2, the optimal dynamic strategy is given by $\theta(t) = -\frac{(\sigma^\top)^{-1}\kappa^\top}{W(t)} \frac{\partial W(t)}{\partial \xi(t)} \xi(t) = -\frac{1-p}{W(t)} \theta^B(t) \frac{\partial W(t)}{\partial \xi(t)} \xi(t)$, where $\theta^B(t)$ stands for the *benchmark investor* strategy defined in [2] as $\theta^B(t) = \frac{1}{1-p}(\sigma^\top)^{-1}\kappa^\top$. Finally, we can define the process $q(t)$ as the exposure to risky assets *relative* to the benchmark portfolio, that we use for further analyses: $\theta^B(t) q(t) = \theta(t)$.

## 4 Numerical Results

To represent a real world market we set the exogenous model parameters as follows: $\alpha = 0.05$, $\delta = 0.15$, $p = -1.5$, $\kappa = 0.4$, $\xi(0) = 1$, $W(0) = 1$, $r = 0.03$ and $T = 1$. In common model applications we observe three market-state intervals in which the portfolio manager behaves differently. In *good* market states (low $\xi_T$) the CVaR portfolio payoff is similar to the benchmark payoff. In *intermediate* states the CVaR portfolio is fully hedged to the level $W_0 - \hat{c}$ and in *the worst* states (high $\xi_T$) the CVaR portfolio is only partially secured. As a result of our hedging strategy we observe an *adjusted* distribution of the terminal portfolio value, indicating lower probability mass concentrated in the left tail, i.e. the probability of attaining the most severe losses is *lower* than that of the benchmark investor (Fig. 1).

In good states the exposition to risky assets is very similar to the benchmark investor. As the market goes down, CVaR investor is selling out risky positions in order to retain the portfolio value above the acceptable level of loss. In the worst cases we observe a *leverage* effect: the investor opens relatively large positions in the risky assets with the intention to rise the portfolio value back to the acceptable level. The portfolio values before investment horizon can be evaluated as a response to the

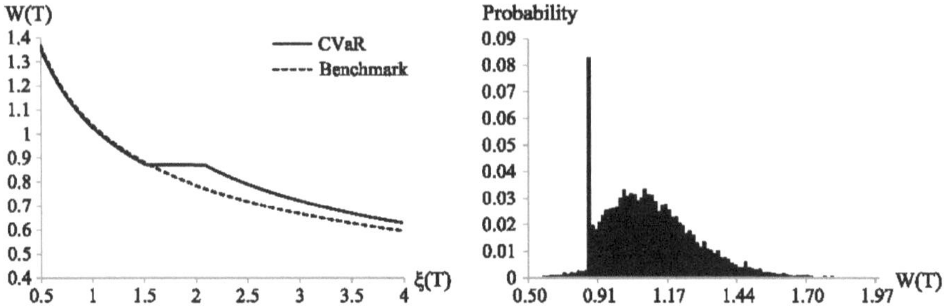


**Fig. 1** *Left*: CVaR terminal portfolio payoff and benchmark terminal portfolio payoff, both as functions of the state variable $\xi(T)$. *Right*: Distribution of the CVaR terminal value payoff for the initial wealth $W_0 = 1$

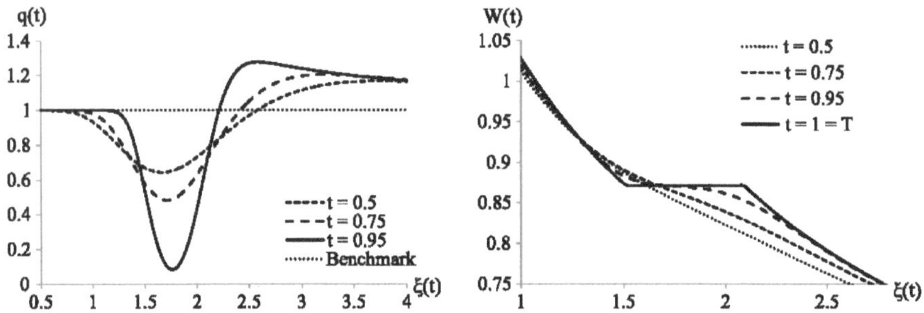


**Fig. 2** *Left*: The dynamics of the relative risky assets exposition as a function of $\xi(t)$. *Right*: CVaR portfolio payoffs convergence to the terminal time shape; as functions of the state variable $\xi(t)$

dynamic investment strategy process $\theta(t)$. As time $t$ approaches $T$, the convergence to the terminal payoff is a necessary condition for the models consistency (Fig. 2).

**Acknowledgements** This work was supported by Vega 1/2429/12 grant.

## References

1. Artzner, P., Delbaen, F., Eber, J.M., Heath, D.: Coherent measures of risk. Math. Finance **9**, 203–228 (1999)
2. Basak, S., Shapiro, A.: Value-at-risk based risk management: optimal policies and asset prices. Rev. Financ. Stud. **14**, 371–405 (2001)
3. Föllmer, H., Schied, A.: Stochastic Finance—An Introduction in Discrete Time. Walter de Gruyter, Berlin (2002)
4. Krokhmal, P., Palmquist, J., Uryasev, S.: Portfolio optimization with conditional value-at-risk objective and constraints. J. Risk **4**(2), 11–27 (2002)

# Pricing and Hedging Variable Annuities

**Abdou Kélani and François Quittard-Pinon**

**Abstract** The aim of this paper is to present a general method to value, hedge and assess risk for a subclass of VA contracts in a Lévy market. This subclass contains Guaranteed Minimum Maturity Benefit (GMMB), Guaranteed Minimum Death Benefit (GMDB), and Guaranteed Minimum Accumulation Benefit (GMAB) that has a cliquet-style option in its design. The suggested unifying method is based on the generalized Fourier transform and gives general quasi-closed form solutions for a large class of Lévy processes.

A numerical analysis that uses a Kou process illustrates the whole procedure.

**Keywords** Variable annuities · Lévy processes

## 1 Products

Let $T$ be the expiration date of the contract, $T_x$ the residual life at time 0 for a policyholder aged $x$ at that time. We denote by $\gamma \leq T$, the time at which the guarantee is triggered. The contract payoffs are of the following type: $\max(F_\gamma, G_\gamma)$, where $F_\gamma$ is the account, or fund value, of the policyholder at time $\gamma$. The quantity $G_\gamma$ is the guarantee that can take many expressions such as a constant equal to the premium, the premium accrued at a guaranteed rate $g$, or the highest account value recorded up to $\gamma$. Because $\max(F_\gamma, G_\gamma)$ is $F_\gamma + [G_\gamma - F_\gamma]^+$, the contract can be considered from the policyholder's point of view as a long position in the fund and a put option with the strike price of $G_\gamma$. The insurer is at risk because of the short position in the option. The policyholder's account value is obtained by continuously deducting at the rate $m$ the value $S$ of the policyholder's portfolio invested in the financial market. We consider $m$ as the sum of two rates: $m_o$ used to fund the guarantee and $m_a$,

A. Kélani · F. Quittard-Pinon (✉)
CEFRA, EMLYON Business School, 23 avenue Guy de Collongue, Écully, France
e-mail: fquittard@wanadoo.fr

A. Kélani
e-mail: abdou.kelani@gmail.com

C. Perna, M. Sibillo (eds.), *Mathematical and Statistical Methods for Actuarial Sciences and Finance*, DOI 10.1007/978-3-319-05014-0_28,

for the other management expenses. In other words, $m = m_o + m_a$. Thus the account value at time $t$ is

$$F_t = F_0 \frac{S_t}{S_0} e^{-mt} = S_t e^{-mt}, \quad F_0 = S_0. \tag{1}$$

To introduce mortality, we denote ${}_t p_x$ as the survival probability $\Pr[T_x > t]$. The probability of dying ${}_t q_x$ is $1 - {}_t p_x$. For convenience, we first consider GMMB and GMDB with a constant guarantee $K$. The insured receives if alive at age $T$, $F_T + [K - F_T]^+$ under the GMMB and $F_{T_x} + [K - F_{T_x}]^+$ under the GMDB. We then consider the GMAB contract which adds the possibility to modify the guarantee at some specified dates by adding a ratchet effect. The investment period $[0, T]$ is subdivided into $T/h$ subintervals of the same duration. Also, we denote as $\xi_{S_0,m_o,t}$ the initial price of the embedded guarantee in the VA contract at time $t$.

## *1.1 Fair Fees and General Valuation Formulas*

Let $r$ be the instantaneous interest rate in the economy, then the discount function is defined by $\delta(t) := \exp\{-\int_0^t r(u)du\}$. The mortality and expense $M\&E$ is the expected discounted value under a risk-neutral measure of all the fees paid until the contract is in force, i.e. until death or maturity, whatever comes first. In an equivalent definition, $M\&E = E_Q[\bar{F}_{T \wedge T_x}]$. Following [2],

$$M\&E(m_o) = F_0 \frac{m_o}{m} \left\{ {}_T p_x \left(1 - e^{-mT}\right) + \sum_{t=0}^{T-1} {}_t p_x q_{x+t} \left(1 - e^{-m(t+1)}\right) \right\}. \tag{2}$$

For a given management rate $m_a$, the equilibrium value or the fair price for the guarantee is the solution in $m_o$ of the equation:

$$\xi_{S_0,m_o,t} = M\&E(m_o). \tag{3}$$

Otherwise stated, the fair cost is such that the value associated with the discounted continuous cash flows coming from the fees is equal to the contract's optional rider value. It is worth noting that this equilibrium equation gives a correspondence between the marginal offset rate $m_o$ and the management rate $m_a$. Also, Eq. (3) takes different expressions according to the particular VA contracts. Using arbitrage pricing theory in continuous time, general formulae can be obtained. The contract values can be expressed as linear combinations of European puts. The GMAB contract incorporates a ratchet effect on the guarantee. To be more precise, a set of rollover dates $\mathcal{A} = \{t_1, t_2, \ldots, t_n\}$, allows that at each of these times the guarantee is modified in the following way: Let $F_t^-$ be the fund value just before the guarantee is reset at time $t$. At time $t_i$, the guarantee becomes: $\max(G_{t_{i-1}}, F_{t_i}^-) = F_{t_i}^- + L_{t_i}$, with $L_{t_i} = [G_{t_{i-1}} - F_{t_i}^-]^+$, where the starting guarantee is $G_0$, and the starting account value is $F_0$. General formulae can be obtained, cf. [2].

## 2 Pricing and Hedging in a Lévy Context

Because of jumps in the financial prices, we are in an incomplete market. In this paper we choose the Esscher risk-neutral measure which makes the discounted prices martingale. For the pricing and hedging, we resort to the generalized Fourier approach in the line of the work of [1]. As shown in Eq. (1), the dynamics of $F$ is the same as those of $S$, up to the continuous dividend rate $q$ equal to $m$. We assume that $S_t = S_0 e^{X_t}$, where $X$ is a Lévy process.

### 2.1 Pricing

The random function $X$, being a Lévy process, can be completely specified by its characteristic exponent, $\psi$: $E(e^{iuX_t}) = e^{-t\psi(u)}$. The price at time $t$ of a European put option with exercise price $K$ and maturity $\tau = T - t$, can be written as

$$P(S,K,q,T) = K\frac{1}{2\pi}e^{-bx'}\int_{\mathbb{R}} e^{iux'}\frac{e^{-\tau(r+\psi_Q(u+ib))}}{(-iu+b)(-iu+b+1)}du, \qquad (4)$$

with $b > 0$, $x' = \ln(S/K)$. It is worth noting that formula (4) holds for a European call option, with $b < -1$. Also note that the expression in Eq. (4) is very suited for a Fast Fourier Transform computation.

### 2.2 Hedging

Using a local risk minimization criterion, the optimal hedging ratio writes:

$$\theta(S_t,T) = K\frac{1}{2S\pi}e^{-bx'}\int_{\mathbb{R}} e^{iux'}\frac{e^{-\tau(r+\psi_Q(u+ib))}B^P(u+ib)}{(-iu+b)(-iu+b+1)}du \quad \text{with } b>0, \qquad (5)$$

where

$$B^P(u) = \frac{-\psi^P(u-i)+\psi^P(u)+\psi^P(-i)}{-\psi^P(-2i)+2\psi^P(-i)}. \qquad (6)$$

The superscripts $P$ and $Q$ refer to parameters in the historical and the risk-neutral worlds, respectively. As can be seen in Eq. (5), the hedging ratio is obtained in exactly the same way the option price is calculated in Eq. (4), up to factors $B^P(.)$ and $S^{-1}$, leading to a unified approach both for pricing and hedging.

## 3 Risk Management

We consider the distribution of future losses which appear as the sum of three components. The first is the discounted value of the residual errors (HE). The second component is given by transaction costs (TC), and the third is the margin amount $\mathcal{M}$ that is deducted by the insurer to fund the guarantee. This amount is also known as the margin offset. Thus, the loss at time $t$ is given by

$$L_t = \mathrm{HE}_t + \mathrm{TC}_t - {}_t p_x \mathcal{M}_t, \tag{7}$$

with $\mathcal{M}_t = m_o \times F_t$, where $m_o$ is the offset ratio obtained via the equilibrium equation (3). Furthermore, the TC are proportional to the absolute variation of the amount allocated in the hedging portfolio: $\mathrm{TC}_t = \mathcal{C}S_t|\Psi_t - \Psi_{t-1}|$, at time $t$. Using Eq. (7) for the risk management of the contracts, we compute the VaR and CTE.

## 4 Illustration

We consider an insured of age 40 years at the contract inception who pays a single 100 USD premium, invested in the referenced portfolio whose initial value is $F_0 = 100$. The interest rate $r$ is assumed to be 6 %. GMMB, GMDB, mixed GMMB/GMDB, and GMAB. The guarantee is assumed to be 100 % and 80 % of the initial premium for the first three contracts and for the last one, respectively. The guarantees within the contract, are funded endogenously at the rate $m_o$, according to the equilibrium equation (3). The hedging strategy uses a MC simulation with 20 000 sample paths and with the $\theta$ ratio. Using a Gompertz-Makeham mortality law and a Kou process for the referenced risky portfolio, we numerically illustrate the whole procedure, see [2]. The following table provides a short summary.

| | Mean | Std | VaR(95 %) | CTE(95 %) |
|---|---|---|---|---|
| GMMB/GMDB: $T = 10$ | −0.2496 | 1.3199 | 1.6965 | 2.2266 |
| GMAB: $\mathcal{A} = \{2, 12, 22\}$ | −13.6641 | 18.3114 | 8.0758 | 14.7923 |

## References

1. Boyarchenko, S., Levendorskiĭ, S.: Option pricing for truncated Lévy processes. Int. J. Theor. Appl. Finance **3**, 549–552 (2000)
2. Kélani, A., Quittard-Pinon, F.: Pricing and hedging variable annuities. Working paper, EM-LYON, May (2013)

# Monetary Risk Functionals on Orlicz Spaces Produced by Set-Valued Risk Maps and Random Measures

**Dimitrios G. Konstantinides and Christos E. Kountzakis**

**Abstract** In this article we study the construction of coherent or convex risk functionals defined either on an Orlicz heart, either on an Orlicz space, with respect to a Young loss function. The Orlicz heart is taken as a subset of $L^0(\Omega, \mathcal{F}, \mu)$ endowed with the pointwise partial ordering. We define set-valued risk maps related to this partial ordering. We also derive monetary risk functionals both by the class of coherent set-valued risk maps defined on them. We also use random measures related to heavy-tailed distributions in order to define monetary risk functionals on Orlicz spaces, whose properties are also compared to the previous ones.

**Keywords** Set-valued risk measure · Random measure · Young loss function · Orlicz heart

## 1 Orlicz Hearts and Orlicz Spaces in Finance

We consider two periods of time (0 and 1) and a non-empty set of states of the world $\Omega$ which is supposed to be an infinite set. The true state $\omega \in \Omega$ that the investors face is contained in some $A \in \mathcal{F}$, where $\mathcal{F}$ is some $\sigma$-algebra of subsets of $\Omega$ which gives the information about the states that may occur at time-period 1. A financial position is a $\mathcal{F}$-measurable random variable $x : \Omega \to \mathbf{R}$. This random variable is the profile of this position at time-period 1. We suppose that the probability of any state of the world to occur is given by a probability measure $\mu : \mathcal{F} \to [0, 1]$. We remind of the use of Orlicz Spaces in Finance, through the notion of *loss function*.

**Definition 1** A function $l : \mathbb{R} \to \mathbb{R}$ is a *loss function* if (a) $l$ is increasing and convex, (b) $l(0) = 0$ and $l(x) \geq x$.

D.G. Konstantinides (✉) · C.E. Kountzakis
Department of Mathematics, University of the Aegean, Karlovassi, 83 200 Samos, Greece
e-mail: konstant@aegean.gr

C.E. Kountzakis
e-mail: chr_koun@aegean.gr

C. Perna, M. Sibillo (eds.), *Mathematical and Statistical Methods for Actuarial Sciences and Finance*, DOI 10.1007/978-3-319-05014-0_29,

Such a function is used to measure the *expected loss* $\mathbb{E}_\mu(l(-x))$ of a financial position $x$. Related to the neutral evaluation of losses $\mathbb{E}_\mu(-x)$ under the state-of-the world measure, the expected loss $\mathbb{E}_\mu(l(-x))$ puts greater weights on high losses than gains. Since $l$ is convex, its convex conjugate $l^*(y) = \sup_{x\in\mathbb{R}}\{xy - l(x)\}$ is well defined (see also [6], p. 4). Let $E$ be a (normed) linear space. A set $C \subseteq E$ satisfying $C + C \subseteq C$ and $\lambda C \subseteq C$ for any $\lambda \in \mathbf{R}_+$ is called *wedge*. A wedge for which $C \cap (-C) = \{0\}$ is called *cone*. A pair $(E, \geq)$ where $E$ is a linear space and $\geq$ is a binary relation on $E$ satisfying the following properties:

(i) $x \geq x$ for any $x \in E$ (reflexive);
(ii) if $x \geq y$ and $y \geq z$ then $x \geq z$, where $x, y, z \in E$ (transitive);
(iii) if $x \geq y$ then $\lambda x \geq \lambda y$ for any $\lambda \in \mathbf{R}_+$ and $x + z \geq y + z$ for any $z \in E$, where $x, y \in E$ (compatible with the linear structure of $E$),

is called *partially ordered linear space*. The partially ordered vector space $E$ is a *vector lattice* if for any $x, y \in E$, the supremum $x \vee y$ and the infimum $x \wedge y$ of $\{x, y\}$ with respect to the partial ordering in $E$. If so, $|x| = \sup\{x, -x\}$ is the *absolute value* of $x$ and if $E$ is also a normed space such that $\| |x| \| = \|x\|$ for any $x \in E$, then $E$ is called *normed lattice*. If a normed lattice is a Banach space, then it is called *Banach lattice*. If we consider the *pointwise partial ordering* on $L^0(\Omega, \mathcal{F}, \mu)$, then the *Orlicz Heart* $X_l := \{x \in L^0 \mid \mathbb{E}_\mu[l(c|x|)] < \infty,\ \forall c > 0\}$ endowed with the *l-Luxemburg* norm

$$\|x\|_l := \inf\left\{a > 0 \mid \mathbb{E}_\mu\left[l\left(\frac{|x|}{a}\right)\right] \leq 1\right\},$$

is a Banach lattice. The norm-dual of $X_l$ is the $X_l^* := \{y \in L^0 \mid \mathbb{E}_\mu[l^*(c|y|)] < \infty, \exists c > 0\}$ Orlicz Space, endowed with the Orlicz norm $\|y\|_{l^*} := \sup\{\mathbb{E}_\mu(xy) : \|x\|_l \leq 1\}$, being equivalent to the $l^*$-Luxemburg norm (see also in [4]). Since $l(x) \geq x$, $x \in \mathbb{R}_+$, then $X_l \subseteq L^1$. We also denote by $M_{l,l^*}(\mu)$ the probability measures on $(\Omega, \mathcal{F})$, which are absolutely continuous with respect to $\mu$ and their densities lie in $X_{l^*}$ (see [6], p. 5). First, we remind the definitions of monetary risk functionals on $X_l$, being alike to the ones met in [2, 5, 7].

**Definition 2** A real-valued function $\rho : X_l \to \mathbb{R}$ is a *coherent monetary risk measure* if it is satisfying

(i) $\rho(x + a\mathbf{1}) = \rho(x) - a$ (Translation Invariance);
(ii) $\rho(\lambda x + (1 - \lambda)x) \leq \lambda\rho(x) + (1 - \lambda)\rho(y)$ for any $\lambda \in [0, 1]$ (Convexity);
(iii) $\rho(\lambda x) = \lambda\rho(x)$ for any $x \in X_l$ and any $\lambda \in \mathbb{R}_+$ (Positive Homogeneity);
(iv) $y \geq x$ implies $\rho(y) \leq \rho(x)$ (Monotonicity),

where $x, y \in X_l$, is called *convex*.

**Definition 3** A correspondence $\rho : X_l \to 2^{X_l}$ is a **coherent risk correspondence** if it satisfies the properties

(i) $\rho(x + a\mathbf{1}) = \rho(x) - \{a\mathbf{1}\}$, $a \in \mathbb{R}$ (Translation Invariance);

(ii) $\rho(\lambda x + (1-\lambda)x) \subseteq \lambda\rho(x) + (1-\lambda)\rho(y)$ for any $\lambda \in [0, 1]$ (Convexity);
(iii) $\rho(\lambda x) = \lambda\rho(x)$ for any $x \in X_l$ and any $\lambda \in \mathbb{R}_+$ (Positive Homogeneity);
(iv) $y \geq x$ implies $\rho(y) \subseteq \rho(x)$ (Monotonicity),

where $x, y \in X_l$.

**Definition 4** A financial position $x$ is **safe with respect to** $\rho : X_l \to 2^{X_l}$, if for this $x \in X_l$, $\rho(x) \subseteq -X_{l,+}$.

**Definition 5** The wedge of the safe financial positions $\mathcal{A}_\rho = \{x \in X_l | \rho(x) \subseteq -X_{l,+}\}$ is called **acceptance set** of $\rho$.

Suppose that $C$ is a wedge of a linear space $E$. A linear functional $f$ of $E$ is called *positive functional* of $C$ if $f(x) \geq 0$ for any $x \in C$. $f$ is a *strictly positive functional* of $C$ if $f(x) > 0$ for any $x \in C \setminus \{0\}$. A linear functional $f$ of $E$, where $E$ is a normed linear space, is called *uniformly monotonic functional* of $C$ if there is some real number $a > 0$ such that $f(x) \geq a\|x\|$ for any $x \in C$.

The set of functions $C^0 = \{f \in E^* | f(x) \geq 0,\ \forall x \in C\}$ is the *dual wedge of C in* $E^*$. If for two wedges $K, C$ of $E$ $K \subseteq C$ holds, then $C^0 \subseteq K^0$. If $C$ is a cone, then a set $B \subseteq C$ is called *base* of $C$ if for any $x \in C \setminus \{0\}$ there exists a unique $\lambda_x > 0$ such that $\lambda_x x \in B$.

The set $B_f = \{x \in C | f(x) = 1\}$, where $f$ is a strictly positive functional of $C$ is the *base of C defined by* $f$. $B_f$ is bounded if and only if $f$ is uniformly monotonic. If $B$ is a bounded base of $C$ such that $0 \notin \overline{B}$ then $C$ is called *well-based*. If $C$ is well-based, then a bounded base of $C$ defined by a $g \in E^*$ exists. Also, $f \in E^*$ is a uniformly monotonic functional of $C$ if and only if $f \in intC^0$, where $intC^0$ denotes the norm-interior of $C^0$.

**Theorem 1** *The* ***derivative monetary coherent risk measure*** $\hat{\rho} : X_l \to \mathbb{R} \cup \{+\infty\}$, *which arises from* $\rho : X_l \to 2^{X_l}$ *is defined as follows*:

$$\hat{\rho}(x) = \sup_{\pi \in \mathcal{A}_\rho^0 \cap M_{l,l^*}(\mu)} \pi(-x), \quad x \in X_l.$$

The consistence range of $\rho$ is used for the definition of a pricing functional $\psi : E \to \mathbb{R}$, which satisfies the properties of a *generalized price* indicated in [1] Lemma 3.3. These properties are listed in the following

**Definition 6** A functional $\psi : E \to \mathbb{R}$ is a **generalized price** if and only if satisfies the following properties:

(i) If $x \geq y$, then $\psi(x) \geq \psi(y)$ ($E_+$-Monotonicity);
(ii) $\psi(x + y) \geq \psi(x) + \psi(y)$ (Super-Additivity);
(iii) $\psi(a \cdot x) = a\psi(x), x \in E, a \in \mathbb{R}_+$ (Positive Homogeneity).

We obtain the following duality between monetary coherent risk measures and generalized prices.

**Theorem 2** *If $\psi : X_l \to \mathbb{R}$ is a* ***generalized price*** *which is normalized at the numeraire financial position $\mathbf{1} \in X_{l,+}$ ($\psi(\mathbf{1}) = 1$) and satisfies the Translation Invariance $\psi(x + a\mathbf{1}) = \psi(x) + a$, $a \in \mathbb{R}$, then $-\psi = \rho_\psi$ is a monetary coherent risk measure. On the other hand if $\rho : X_l \to \mathbb{R}$ is a monetary coherent risk measure, then $-\rho = \psi_\rho : E \to \mathbb{R}$ is a generalized price functional.*

## 2 Monetary Risk Functionals Arising from Random Measures

We consider the probability space $(\Omega, \mathcal{F}, \mu)$ and a measurable space $(E, \mathcal{E})$. A *random measure* (see [8]) $\xi$ on $(E, \mathcal{E})$ over the probability space $(\Omega, \mathcal{F}, \mu)$ is a map $\xi : \mathcal{E} \times \Omega \to \mathbb{R}_+$, such that

1. the map $\omega \mapsto \xi(A, \omega)$ is a random variable for any $A \in \mathcal{E}$;
2. the map $A \mapsto \xi(A, \omega)$ is a measure on $\mathcal{E}$, $\mu$-almost surely in $\Omega$.

The measurable $(E, \mathcal{E})$ is actually $(S, \mathcal{B}_S)$, where $\mathcal{B}$ is the Borel $\sigma$-algebra on $S$. If we would like to create a $\mathcal{V}$-related coherent $\rho : X_l \to \mathbb{R}$, where $\mathcal{V}$ is a family of heavy-tailed distributions (see [3]), such that the densities of $\xi(y)$, $y \in S$ belong to $M_{l,l^*}(\mu)$, where $S \subseteq Y$ is compact and $Y$ is a finite-dimensional topological manifold $Y$ (see [9]) of parameters of $\mathcal{V}$.

**Theorem 3** *The following risk measure is coherent on $X_l$*

$$\mathbb{ES}_{a,\mathcal{V},S,l}(x) = \sup_{0 \leq \frac{d\mu_{\xi(y)}}{d\mu} \leq \frac{1}{a},\ y \in S} \mathbb{E}_{\mu_{\xi(y)}}(-x), \quad x \in X_l.$$

## References


1. Aliprantis, C.D., Tourky, R., Yannelis, N.C.: A theory of value with non-linear prices: equilibrium analysis beyond vector lattices. J. Econ. Theory **100**, 22–72 (2001)
2. Artzner, P., Delbaen, F., Eber, J.M., Heath, D.: Coherent measures of risk. Math. Finance **9**, 203–228 (1999)
3. Cai, J., Tang, Q.: On max-sum equivalence and convolution closure of heavy-tailed distributions and their applications. J. Appl. Probab. **41**, 117–130 (2004)
4. Cheridito, P., Li, T.: Risk measures on Orlicz hearts. Math. Finance **19**, 189–214 (2009)
5. Delbaen, F.: Coherent risk measures on general probability spaces. In: Advances in Finance and Stochastics: Essays in Honour of Dieter Sondermann, pp. 1–38. Springer, Berlin (2002)
6. Drapeau, S., Kupper, M., Papapantoleon, A.: A Fourier Approach to the Computation of CV@R and Optimized Certainty Equivalents (2013)
7. Föllmer, H., Schied, A.: Convex measures of risk and trading constraints. Finance Stoch. **6**, 429–447 (2002)
8. Karr, A.: Lévy random measures. Ann. Probab. **6**, 57–71 (1978)
9. Lee, J.M.: Introduction to Topological Manifolds. Graduate Texts in Mathematics, vol. 202. Springer, New York (2000)

# A Probability Inequality Related to Mardia's Kurtosis

**Nicola Loperfido**

**Abstract** We use the measure of multivariate kurtosis introduced by Mardia to define an upper bound for the probability that the Mahalanobis distance of a random vector from its mean is greater or equal than a given value. The bound improves on a similar one, based on Markov's theorem, and generalizes to the multivariate case an inequality which appears in several textbooks. It might be applied whenever the distribution of the Mahalanobis distance of a random vector from its mean is not easily computable, as it is often the case in finance and actuarial sciences.

**Keywords** Kurtosis · Mahalanobis distance · Markov's theorem

## 1 Introduction

Let $x$ be a $d$-dimensional random vector with mean $\mu$, nonsingular covariance matrix $\Sigma$ and finite fourth-order moments. Mardia [8] proposed to measure the kurtosis of $x$ by

$$\beta_{2,d}(x) = E\left[(x-\mu)^T \Sigma^{-1}(x-\mu)\right]^2. \tag{1}$$

In the univariate case, (1) coincides with Pearson's kurtosis, that is the fourth moment of a standardized random variable. Mardia's kurtosis has become the best known and most used measure of multivariate kurtosis [4, 6, 14]. Mardia [8] applied Markov's theorem to obtain

$$P\left[(x-\mu)^T \Sigma^{-1}(x-\mu) \geq \varepsilon\right] \leq \frac{\beta_{2,d}}{\varepsilon^2}, \tag{2}$$

which is useless when $\varepsilon \leq d$, since $\beta_{2,d} \geq d^2$ [8]. For a random variable $X$ with mean $\mu$, standard deviation $\sigma$ and standardized fourth moment $\beta_2$, it becomes

$$P\left(\left|\frac{X-\mu}{\sigma}\right| \geq \lambda\right) \leq \frac{\beta_2}{\lambda^4}, \tag{3}$$

N. Loperfido (✉)
Università degli Studi di Urbino "Carlo Bo", via Saffi 42, 61029 Urbino (PU), Italy
e-mail: nicola.loperfido@uniurb.it

C. Perna, M. Sibillo (eds.), *Mathematical and Statistical Methods for Actuarial Sciences and Finance*, DOI 10.1007/978-3-319-05014-0_30,

which can be improved as follows, for $\lambda > 1$:

$$P\left(\left|\frac{X-\mu}{\sigma}\right| \geq \lambda\right) \leq \frac{\beta_2 - 1}{(\lambda^2-1)^2 + \beta_2 - 1}. \tag{4}$$

The above inequality appears in several textbooks: [2] (p. 256), [13] (p. 102), [12] (pp. 53–55) and [11] (p. 323), where it is introduced as the David and Barton inequality. The present paper sharpens (2) by generalizing (4) to the multivariate case. The next section contains the main result and the last one comments it.

## 2 Inequality

This section states the inequality and proves it. The inequality is related to Markov's theorem, which appears in its proof. Probability inequalities related to Markov's theorem are thoroughly discussed by [3].

**Theorem 1** *Let $\beta_{2,d}$ be Mardia's kurtosis of a random vector $x$ with mean $\mu$ and nonsingular covariance $\Sigma$. Then the following inequalities hold for any real $\varepsilon$ greater than $d$:*

$$P\left[(x-\mu)^T \Sigma^{-1}(x-\mu) \geq \varepsilon\right] \leq \frac{\beta_{2,d} - d^2}{\varepsilon^2 - 2d\varepsilon + \beta_{2,d}} \leq \frac{\beta_{2,d}}{\varepsilon^2}. \tag{5}$$

*Proof* Let $Y = (x-\mu)^T \Sigma^{-1}(x-\mu)$ be the squared Mahalanobis distance of $x$ from its mean $\mu$ and consider the mean of $[(Y-d)(\varepsilon-d) + (\beta_{2,d} - d^2)]^2$, that is

$$(\varepsilon-d)^2 E\left[(Y-d)\right]^2 + 2(\varepsilon-d)\left(\beta_{2,d} - d^2\right)E(Y-d) + \left(\beta_{2,d} - d^2\right)^2. \tag{6}$$

The above expectation might be simplified into

$$(\varepsilon-d)^2\left(\beta_{2,d} - d^2\right) + \left(\beta_{2,d} - d^2\right)^2 = \left(\beta_{2,d} - d^2\right)\left(\varepsilon^2 - 2d\varepsilon + \beta_{2,d}\right) \tag{7}$$

by noticing that $E(Y) = d$ and $E(Y^2) = \beta_{2,d}$. It follows that the expectation of the random variable

$$W = \left[\frac{(Y-d)(\varepsilon-d) + (\beta_{2,d} - d^2)}{\varepsilon^2 - 2d\varepsilon + \beta_{2,d}}\right]^2 \tag{8}$$

is $(\beta_{2,d} - d^2)/(\varepsilon^2 - 2d\varepsilon + \beta_{2,d})$. Since $W$ is nonnegative with finite expectation Markov's theorem leads to

$$P(W \geq 1) \leq \frac{\beta_{2,d} - d^2}{\varepsilon^2 - 2d\varepsilon + \beta_{2,d}}. \tag{9}$$

The assumption $\varepsilon > d$ implies that $P(W \geq 1) = P(Y \geq \varepsilon)$, as it can be seen from the inequalities $(Y-d)(\varepsilon-d) + (\beta_{2,d} - d^2) \geq \varepsilon^2 - 2d\varepsilon + \beta_{2,d}$, $(Y-d)(\varepsilon-d) \geq$

$\varepsilon^2 - 2d\varepsilon + d^2$ and $(Y - d) \geq (\varepsilon - d)$. We can now complete the first part of the proof by recalling the definition of $Y$:

$$P\left[(x-\mu)^T \Sigma^{-1}(x-\mu) \geq \varepsilon\right] \leq \frac{\beta_{2,d} - d^2}{\varepsilon^2 - 2d\varepsilon + \beta_{2,d}}. \tag{10}$$

We shall now prove the second part of the theorem. A little algebra shows that the inequality

$$\frac{\beta_{2,d}}{\varepsilon^2} \geq \frac{\beta_{2,d} - d^2}{\varepsilon^2 - 2d\varepsilon + \beta_{2,d}} \tag{11}$$

is equivalent to

$$\frac{(\beta_{2,d} - d\varepsilon)^2}{\varepsilon^2[(\varepsilon - d)^2 + \beta_{2,d} - d^2]} \geq 0, \tag{12}$$

which always holds since $\beta_{2,d} \geq d^2$ [8] and $\varepsilon > d$. This completes the proof. □

# 3 Remarks

The following remarks are meant to give a better insight into the previous section's inequality.

The distribution of the Mahalanobis distance of a random vector from its mean has a simple analytical form only in a few cases, notably the normal and $t$ ones. In the general case it is not easily computable: finance and actuarial sciences provide many such examples. On the contrary, Mardia's kurtosis has a simple analytical form for many well-known multivariate distributions: normal [8], mixture of two normals [9], bivariate uniform, bivariate Pareto, bivariate exponential [10], scale mixture of skew-normal distributions ([5]), multivariate Laplace [7], elliptical distributions [14], multivariate $t$, multivariate Pearson type II, multivariate Pearson type VII, multivariate symmetric Kotz type [1]. Hence Theorem 1 might be helpful in approximating the cumulative distribution function of the Mahalanobis distance of a random vector from its mean.

The assumption $\varepsilon > d$ cannot removed without replacing it with other assumptions, as shown in the following example. Let $X_1, \ldots, X_d$ be pairwise independent random variables satisfying $P(X_i = 1) = P(X_i = -1) = 0.5$, for $i = 1, \ldots, d$. It easily follows that the mean and the variance of $x = (X_1, \ldots, X_d)^T$ equal the null vector and the identity matrix, respectively. Moreover, Mardia's kurtosis of $x$ attains its minimum value, that is $d^2$. The Mahalanobis distance of $x$ from its mean is $d$ with probability one. The upper bound of the above inequality is zero, but the probability of the Mahalanobis distance of $x$ from its mean being greater or equal to $d$ is one. Hence the above inequality does not always hold when $\varepsilon$ is not constrained to be greater than the vector's dimension.

**Acknowledgements** The Author wish to thank Prof. Domenico Piccolo for the encouragement.

## References


1. Baringhaus, L., Henze, N.: Limit distributions for Mardia's measure of multivariate skewness. Ann. Stat. **20**, 1889–1902 (1992)
2. Cramér, H.: Mathematical Methods of Statistics. Princeton University Press, Princeton (1946)
3. Ghosh, B.K.: Probability inequalities related to Markov's theorem. Am. Stat. **56**, 186–190 (2002)
4. Henze, N.: On Mardia's kurtosis test for multivariate normality. Commun. Stat., Theory Methods **23**, 1031–1045 (1994)
5. Kim, H.M.: A note on scale mixtures of skew-normal distributions. Stat. Probab. Lett. **78**, 1694–1701 (2008)
6. Kollo, T.: Multivariate skewness and kurtosis measures with an application in ICA. J. Multivar. Anal. **99**, 2328–2338 (2008)
7. Kollo, T., Srivastava, M.S.: Estimation and testing of parameters in multivariate Laplace distribution. Commun. Stat., Theory Methods **33**, 2363–2687 (2004)
8. Mardia, K.V.: Measures of multivariate skewness and kurtosis with applications. Biometrika **57**, 519–530 (1970)
9. Mardia, K.V.: Applications of some measures of multivariate skewness and kurtosis in testing normality and robustness studies. Sankhya, Ser. B **36**, 115–128 (1974)
10. Mardia, K.V.: Assessment of multinormality and the robustness of Hotelling's $T^2$ test. J. R. Stat. Soc., Ser. C, Appl. Stat. **24**, 163–171 (1975)
11. Piccolo, D.: Statistica. Il Mulino, Bologna (2010)
12. Pieraccini, L.: Fondamenti di Inferenza Statistica, 2nd edn. Giappichelli, Torino (2007)
13. Rohatgi, V.K.: An Introduction to Probability Theory and Mathematical Statistics. Wiley, New York (1976)
14. Zografos, K.: On Mardia's and Song's measures of kurtosis in elliptical distributions. J. Multivar. Anal. **99**, 858–879 (2008)

# Integrating Industrial and Financial Analysis into a Rating Methodology for Corporate Risk Detection: The Case of the Vicenza Manufacturing Firms

**Guido Max Mantovani, Giancarlo Coro, Paolo Gurisatti, and Mattia Mestroni**

**Abstract** Banks weakness derived from rating models that produce cyclical effects on credit availability and are not able to anticipate anti-cyclical firms' trends. The aim of the paper is to develop a framework for an original rating methodology derived from integration of industrial and financial analysis able to identify best performers in crisis scenarios (anti-cyclically). Industrial analysis is based on firm heterogeneity approaches to measure three dimensions of analysis: innovation, internationalization and growth. Financial analysis focuses on operational return and risks measures and develops an integrated classification of firms using standardized XBRL financial data. Further integration of the two methodologies is used to create the effective set of information needed for rating system.

**Keywords** Corporate risks · Rating · Firm behavior · Firm performance

## 1 Introduction

Techniques adopted in the classical financial risk approaches are often useless, since they are based on the heterogeneous nature of risks. Instead, in real terms corporate risks have huge endogenous components. Risk is continuously crafted by managerial decisions, including those adopted in order to manage them. The simple financial approach in corporate risk management is reductive, missing the business

G.M. Mantovani (✉)
HERMES-Universities, 61 Avenue de la Foret Noire, 67085 Strasbourg Cedex, France
e-mail: president@hermes-universities.eu

G. Coro · P. Gurisatti · M. Mestroni
Ca' Foscari University of Venice, Fondamenta San Giobbe 873, 30121 Venezia, Italy

G. Coro
e-mail: corog@unive.it

P. Gurisatti
e-mail: paolo.gurisatti@unive.it

M. Mestroni
e-mail: mattia.mestroni@unive.it

C. Perna, M. Sibillo (eds.), *Mathematical and Statistical Methods for Actuarial Sciences and Finance*, DOI 10.1007/978-3-319-05014-0_31,

model determinants along with the managerial decisions contribution. An integrated approach is then required, in order to soundly support the managerial choices.

## 2 Literature Review

The great financial crisis of 2008 has shown all the weaknesses of a World Bank regulation that presents high levels of pro-cyclical effects [3, 4]. The evidence of such limits and threats incorporated in Basel regulation was extensively proved by academic world [5, 8, 10]. The high proportion of SMEs and the high productivity of North East [1, 7] drive our choices on manufacturing firms of Vicenza to test our original rating methodology. The integration of financial methodology with an industrial one puts our work in line with some precedent papers that underline the importance of adding soft information to standard financial approaches to a correct valuation of firms credit merit [4, 6].

## 3 Industrial Analysis

This investigation was initiated with a survey to a sample of 309 industrial firms, selected by industry and size representativeness, located in Vicenza. Our research hypothesis refers to firm heterogeneity approaches [2]: firstly we look at innovation capabilities, collecting data on patents and R&D offices; secondly, we evaluate the international activities through information on firm's export, the occurrence of affiliates abroad and where firms' main competitors are; finally, we measured the turnover and profit performance just after the 2008 crisis. According with the emerging characteristics, we named the five groups as following:

- G1–International and reactive firm (about 20 per cent of the sample);
- G2—International but not reactive firm (15 per cent of the sample);
- G3—Local reactive firm (20 per cent of the sample);
- G4—National or local not reactive firms (15 per cent of the sample);
- G5—Average or standard firms (30 per cent of the sample).

## 4 The Sample

Any authority identification code of the 309 firms considered in industrial analysis was available in our data set of manufacturing firms located in Vicenza. Using identification codes, by AIDA database research function we could extract complete balance sheets for financial analysis implementation. The analysis was performed on a sample data containing continuous and complete 2004–2012 standard financial reports.

**Table 1** Financial model framework for the classification of firms (ROC = Return On Capital)

| | | RETURNS | | |
|---|---|---|---|---|
| | | INCREASING ROC ($ROC_{2010} > ROC_{2007}$) | | DECREASING ROC ($ROC_{2010} < ROC_{2007}$) |
| | | STEADY TREND ($ROC_{2010} > ROC_{2008}$) | UNSTEADY TREND ($ROC_{2010} < ROC_{2008}$) | WORSENING |
| RISK | DECREASING RISK | OK | Anomalous Firms to be reclassified | Critic firms |
| | INCREASING RISK | Anomalous Firms to be reclassified | Critic firms | KO |

## 5 Financial Analysis

The intuition behind this model is the need to give an appropriate emphasis to risk dimensions in classifying a "performing" firm. The resulting matrix (Table 1) classifies the sample into six quadrants: "OK firms", "KO firms", two quadrants identified as "Critic Firms" and two quadrants identified as "Anomalous Firms to be reclassified". Three risk dimensions considered are Degree of Operative Leverage (DOL) for both price and volumes changes in operating revenue, and working capital absolute intensity, that is the working capital on operating revenue rate. Risk indexes and ROC definitions follow previous research standards defined by [9]. The above analysis is performed over three timeframes: the pre-crisis period (2004–2007); the crisis period (2007–2010) and the post-crisis period (2010–2012).

## 6 Results

In sum, we can say that industrial analysis produce a consistent method to identify best performers, confirmed by post crisis financial analysis. On one hand, the industrial method identifies firms with high return rate and low risk exposure—as G1, G3 and G5—and firms with low return rate and high risk exposure—as G2 and G4. On the other, the financial method confirm the capacity to react to crisis of best performers groups, identified by industrial methodology—G1 and G3—and the expectations about cluster performance are confirmed also after crisis.

## 7 Conclusions

Financial analysis demonstrate that industrial classification identifies correctly the cluster G1 as best performers: it has the best capacity to react to crisis and the high percentage of OK firms during the pre-crisis, the crisis and the post-crisis periods. Also G4 was correctly identified as the worst performer group: it has the lowest percentage of improving firms and the highest percentage of worsening firms in crisis and post crisis timeframes. As industrial model predicts, G3 and G5 result as clusters of good performers even if due to different features. The most interesting cluster is the G2 group, defined by the industrial analysis as a group of international players firms with low performance. During crisis, this group suffers a high degree of risk exposure that is the reason of the low performance on the three timeframes. But, its international openness permitted to reduce risk exposure after crisis and G2 report an improvement in financial classification, even if conserving low ROC levels. The results of the empirical analysis are clear: an integrated approach in corporate risk detection is clearly more efficient. By adopting such a methodology you must measure the impact of risks that do persist into the firm, along with their impact as a bundle.

## References

1. Bank of Italy: The economy of the North East. Workshops and Conferences **2011**(8), 1–759 (2011)
2. Bernard, A.B., Redding, S.J., Schott, P.K.: The empirics of firm heterogeneity and international trade. Annu. Rev. Econ. **4**(1), 283–313 (2012)
3. Blundell-Wignall, A., Atkinson, P.: Thinking beyond Basel III: necessary solutions for capital and liquidity. OECD J., Financ. Mark. Trends **2010**(1), 9–33 (2010)
4. Dainelli, F., Giunta, F., Cipollini, F.: Determinants of SME credit worthiness under Basel rules: the value of credit history information. PSL Q. Rev. **66**(264), 21–47 (2013)
5. Kashyap, S.: Cyclical implications of the Basel-II capital standards. FRB Chic. Econ. Perspect. **28**(1), 18–33 (2003)
6. Liberti, J.M.: How does organizational form matter? Distance, communication and soft information. London Business School, Mimeo (2005)
7. Manfra, P.: Entrepreneurship, firm size and the structure of the Italian economy. J. Entrep. Finance **7**(3), 99–111 (2002)
8. Mantovani, G.M., Daniotti, E.: Valori e Capitali per un Nuovo Patto di Sviluppo del Sistema: Il Caso Treviso. Edizioni Ca' Foscari—Digital Publishing, Venice (2012)
9. Mantovani, G.M., Daniotti, E., Gurisatti, P.: In search of corporate risk measures to complete financial reporting. The case of the "Caldarerie" industry. Int. Res. J. Appl. Finance **4**(3), 458–489 (2013)
10. Sironi, A., Zazzara, C.: The new Basel accord: implications for Italian banks. Rev. Financ. Econ. **12**, 99–126 (2003)

# Risk Measurement Using the Mixed Tempered Stable Distribution

**Lorenzo Mercuri and Edit Rroji**

**Abstract** The Mixed Tempered Stable distribution (MixedTS) recently introduced has as special cases parametric distributions used in asset return modelling such as the Variance Gamma (VG) and Tempered Stable. In this paper, we start from this flexible distribution and compare the historical estimates for the two homogeneous risk measures with the quantities obtained using direct numerical integration and the saddle-point approximation. The homogeneity property enables us to go further and look for the most important sources of risk. Although risk decomposition in a parametric context is not straightforward, modified versions of VaR and ES based on asymptotic expansions simplify the problem.

**Keywords** MixedTS · Homogeneity · Risk decomposition

## 1 Model Description

Value at risk [6] and Expected Shortfall [10] have emerged as industry standards for measuring downside risk. Non-parametric approaches for their estimation have gained the consensus of practitioners since they are easier to implement and only the information in the return series is used. However, non-parametric methods imply more uncertain estimates for the risk measures considered as shown for example in [1]. In this paper we consider a parametric distribution, the Mixed Tempered Stable [9], for modeling asset returns and use it in risk computation. We say that a continuous random variable $Y$ follows a Mixed Tempered Stable distribution if:

$$Y \stackrel{d}{=} \sqrt{V}\tilde{X} \tag{1}$$

L. Mercuri
University of Milan, via del Conservatorio 7, Milan, Italy
e-mail: lorenzo.mercuri@unimi.it

E. Rroji (✉)
University of Milano-Bicocca, via degli Arcimboldi 8, Milan, Italy
e-mail: edit.rroji@unimib.it

C. Perna, M. Sibillo (eds.), *Mathematical and Statistical Methods for Actuarial Sciences and Finance*, DOI 10.1007/978-3-319-05014-0_32,

where $\tilde{X}|V \sim stdCTS(\alpha, \lambda_+\sqrt{V}, \lambda_-\sqrt{V})$. $V$ is a Lévy distribution defined on positive axis and its m.g.f always exists. The logarithm of the m.g.f. is:

$$\Phi_V(u) = \ln\big[E\big[\exp(uV)\big]\big]. \tag{2}$$

We compute the characteristic exponent for the new distribution and apply the law of iterated expectation:

$$\begin{aligned} E\big[e^{iu\sqrt{V}\tilde{X}}\big] &= E\big\{E\big[e^{iu\sqrt{V}\tilde{X}}|V\big]\big\} \\ &= \exp\big[\Phi_V\big(L_{stdCTS}(u;\alpha,\lambda_+,\lambda_-)\big)\big]. \end{aligned} \tag{3}$$

The characteristic function identifies the distribution at time one of a time changed Lévy process and the distribution is infinitely divisible.

**Proposition 1** *The first four moments for the MixedTS have an analytic expression*:

$$\begin{cases} E[\sqrt{V}\tilde{X}] = 0 \\ Var[\sqrt{V}\tilde{X}] = E[V] \\ \gamma_1 = (2-\alpha)\dfrac{(\lambda_+^{\alpha-3} - \lambda_-^{\alpha-3})}{(\lambda_+^{\alpha-2} + \lambda_-^{\alpha-2})} E^{-1/2}[V] \\ \gamma_2 = \left[3 + (3-\alpha)(2-\alpha)\dfrac{(\lambda_+^{\alpha-4} + \lambda_-^{\alpha-4})}{(\lambda_+^{\alpha-2} + \lambda_-^{\alpha-2})}\right] \dfrac{E[V^2]}{E^2[V]}. \end{cases} \tag{4}$$

The choice of using this distribution comes from the fact that if we assume that $V \sim \Gamma(a, \sigma^2)$, we have as special cases some well-known distributions in modeling financial returns. We get the VG [8] for $\alpha = 2$ and the standardized Classical Tempered Stable [4] (*stdCTS*) when $\sigma = \frac{1}{\sqrt{a}}$ and $a$ goes to infinity.

Edgeworth expansions are frequently used to approximate distributions when higher order moments are available. This methodology seems to work well in the center of the distribution but it often produces negative values for densities in the tails. Saddle-point expansion [7] can be understood as an refinement of the Edgeworth expansion on the tails. In [3] is given a general description about how to approximate the density and the cumulative distribution function (cdf) of a continuous r.v. $X$ whose moment generating function (mgf) $M_X(t)$ exists in an open set around zero. Given the cumulant generating function (cgf)

$$K_X(t) = \ln\big[M_X(t)\big] \tag{5}$$

the solution $\hat{s} = s(x)$ of the equation $x = K'_X(\hat{s})$ is the saddle-point at $x$ belonging to the support of the r.v $X$. The first order Saddle-Point approximation for the density $f_X(x)$ is:

$$\hat{f}_X(x) = \frac{1}{\sqrt{2\pi K''_X(\hat{s})}} \exp\big\{K_X(\hat{s}) - x\hat{s}\big\} \tag{6}$$

while continuing with the derivation is it is possible to arrive at the second order approximation and the existence of the first four moments becomes crucial. After

**Table 1** Comparison of the Expected Shortfall at level 5 % using the empirical distribution and the parametric MixedTS distribution. The dataset is compose by daily log returns ranging from 14-June-2011 to 20-September-2013 of the VFIAX, which is a fund that tries to replicate the performance of the S&P 500 index, and the ten sector indexes in the USA market

| | $ES_{0.05}^{MixedTS}$ | $ES_{0.05}^{Fourier}$ | $ES_{0.05}^{SPA1}$ | $ES_{0.05}^{SPA2}$ | $ES_{0.05}^{Emp}$ |
|---|---|---|---|---|---|
| VFIAX | 0.0173 | 0.0179 | 0.0205 | 0.0204 | 0.0170 |
| COND | 0.0157 | 0.0158 | 0.0155 | 0.0164 | 0.0170 |
| CONS | 0.0176 | 0.0177 | 0.0200 | 0.0196 | 0.0163 |
| ENRS | 0.0184 | 0.0185 | 0.0182 | 0.0192 | 0.0230 |
| FINL | 0.0192 | 0.0194 | 0.0194 | 0.0202 | 0.0206 |
| HLTH | 0.0177 | 0.0188 | 0.0209 | 0.0204 | 0.0171 |
| INDU | 0.0210 | 0.0210 | 0.0235 | 0.0233 | 0.0197 |
| INFT | 0.0209 | 0.0206 | 0.0231 | 0.0223 | 0.0191 |
| MATR | 0.0215 | 0.0211 | 0.0212 | 0.0218 | 0.0233 |
| TELS | 0.0186 | 0.0188 | 0.0189 | 0.0192 | 0.0216 |
| UTIL | 0.0161 | 0.0161 | 0.0165 | 0.0165 | 0.0170 |

testing the ability of the new distribution to reproduce the features of the returns observed in the market, we focus on the computation of risk measures in a parametric context in order to exploit the results obtained previously. As a first exercise, we fit the MixedTS directly to the returns of Vanguard Fund Index which tries to replicate the S&P 500 and to the ten indexes. The fitted parameters are used for the computation of the risk measures.
We evaluate the cumulative distribution function using the formula that through the characteristic function $\phi_X(t)$ allows us to evaluate the cdf $F_X(x)$:

$$F_X(x) = \frac{1}{2} - \frac{1}{2\pi}\int_{-\infty}^{+\infty} \frac{[e^{-itx}\phi_X(t)]}{it}\mathrm{d}t. \tag{7}$$

We consider the VaR computed using the Inverse Fourier Transform and compare it with that obtained with the historical simulation approach. In Table 1, for each index, we compare the empirical ES estimates with the estimates using four different methodologies.

1. Numerical integration of the expected value, i.e the distribution is obtained through the Inverse Fourier Transform. We recall that, under the assumption of the existence for the $E(X)$,[1] the Expected Shortfall can be written as

$$ES_\alpha(X) := E[X|X \leq x_\alpha] = x_\alpha - \frac{1}{\alpha}\int_{-\infty}^{x_\alpha} F(u)\mathrm{d}u; \tag{8}$$

2. The second approach is based on Monte Carlo simulation. The random number generator is built using the Inverse Transform Sampling method [5];

[1] For the Mixed Tempered Stable distribution, this condition is ensured by the existence of moment generating function.

3. First order Saddle Point Approximation formula.

$$ES_\alpha(X) = \frac{1}{\alpha}\int_{-\infty}^{q} x f_X(x)dx \approx \frac{1}{\alpha}\left[\mu_X F_X(q) - f_X(q)\frac{q-\mu_X}{\hat{s}}\right]; \tag{9}$$

4. Second order Saddle Point Approximation formula.

$$\hat{ES}_\alpha(X) = \Phi(\hat{\omega}_q)\mu_X + \sqrt{\frac{1}{2\pi}}\exp\left\{-\frac{\hat{\omega}_q^2}{2}\right\}\left[\frac{\mu_X}{\hat{\omega}_q} - \frac{q}{\hat{u}_q}\right] \tag{10}$$

where $\hat{\omega}_q$, $\hat{u}_q$ are evaluated in the quantile point $x = q$.

The study of risk attribution using the MixedTS is based on the definitions of modified VaR [11] and modified ES [2] that consider asymptotic expansions and require only the existence of the first four moments. Parametric risk decomposition is more difficult than a simple historical approach but estimates are more stable and accurate.

## References

1. Aussenegg, W., Miazhynskaia, T.: Uncertainty in value-at-risk estimates under parametric and non-parametric modeling. Financ. Mark. Portf. Manag. **20**(3), 243–264 (2006)
2. Boudt, K., Peterson, B.H., Croux, C.: Estimation and decomposition of downside risk for portfolios with non-normal returns. J. Risk **11**(2), 79–103 (2009)
3. Cizek, P.J., Hardle, W., Weron, R.: Statistical Tools for Finance and Insurance, 2nd edn. Springer, Berlin (2011)
4. Cont, R., Tankov, P.: Financial Modelling with Jump Processes, 2nd edn. Chapman & Hall/CRC Financial Mathematics Series (2003)
5. Glasserman, P.: Monte Carlo Methods in Financial Engineering, 1st edn. Stochastic Modelling and Applied Probability. Springer, Berlin (2003)
6. Longerstaey, J., Zangari, P.: RiskMetrics: Technical Document, 4th edn. Morgan Guaranty Trust Co., New York (1996)
7. Lugannani, R., Rice, S.: Saddlepoint approximation for the distribution of the sum of independent random variables. Adv. Appl. Probab. **12**, 475–490 (1980)
8. Madan, D.B., Seneta, E.: The variance gamma (V.G.) model for share market returns. J. Bus. **63**, 511–524 (1990)
9. Rroji, E.: Risk decomposition and semi heavy tailed distributions. PhD Thesis, Univ. Milano-Bicocca (2013)
10. Tasche, D.: Expected shortfall and beyond. J. Bank. Finance **26**(7), 1519–1533 (2002)
11. Zangari, P.: A VaR methodology for portfolios that include options. RiskMetrics Monit. **7**, 4–12 (1996)

# Corporate Finance… What Else? The Case of the Productive Chain Networks in North-East Italy and the Scaffolding Finance Adopted by Their Leader

**Mattia Mestroni, Elisabetta Basilico, and Guido Max Mantovani**

**Abstract** The Italian North-East district is a clear evidence of the presence of productive chain networks, in which firms tend to specialize in specific risk management. This generates new approaches in the theory of the firm. We investigate which are the implications for the financial activities of the clusters. The paper presents a methodology to identify firms according to their network role: LF, SF and standing alone firms (SA). Accordingly, empirical evidence about the capital raising activity of LF, SF and SA is reported, deploying the necessity of a new approach in finance, where "corporation" is no more the focus.

**Keywords** Networks · Firm boundaries · Corporate finance · Working capital

## 1 Introduction

Can we still adopt the expression "corporate finance" in an economic environment where the detection of the boundaries of corporations is becoming more and more difficult? Maybe the answer is "no more"! The necessity to satisfy the increasingly evolving and unique needs, together with the necessity to specialize in the use of productive factors, requires a continuous evolution of the firm concept. Barney and Ouchi [1] collected a wide range of papers, explaining why some transactions can be better managed by approaches that stand-in-the-middle between pure markets and firm organizations. Within this framework, and as [2] predicted, large corporations are generated only when the firm organization can be more efficient than markets. In other cases, markets can be more efficient, but in some others, hybrids models are

M. Mestroni (✉) · E. Basilico
Ca' Foscari University of Venice, Fondamenta San Giobbe 873, 30121 Venezia, Italy
e-mail: mattia.mestroni@unive.it

E. Basilico
e-mail: elisabetta.basilico@unive.it

G.M. Mantovani
HERMES-Universities, 61 Avenue de la Foret Noire, 67085 Strasbourg Cedex, France
e-mail: president@hermes-universities.eu

C. Perna, M. Sibillo (eds.), *Mathematical and Statistical Methods for Actuarial Sciences and Finance*, DOI 10.1007/978-3-319-05014-0_33,

required: transactions are still managed by markets and they are assisted by a clan (i.e. semi-organizational) agreement [1]. Small and Medium Enterprise (SME) businesses give strong empirical insights about this evidence of the evolution toward a twin-necessities-satisfaction. Very challenging SMEs are now competing more and more, by acting together into clusters that coordinate their actions [6]. Sometimes, this cluster competition is mainly driven by the nature of the sold products (e.g. in the case of chains of firms); in other cases, technology is the main driver of the competitive advantage of the cluster (e.g. in the case of districts). In both cases the cluster acts similarly to a unique firm, but its organization nexus is based on market transactions, assisted by clan's rules. This approach allows any single member of the cluster, to specialize in the areas where its own skills are efficient at most. Two possible approaches can be configured: (i) a network/partnership agreement, which is the institution where the money is flowed to; (ii) a leader (in finance), which is identified inside the cluster and appointed as financial manager for the entire group. Clusters already self-appoint leaders to manage their transaction (i.e. main contractors for selling activities, productive leader to spread skills, etc.). The mission-critical role of such leaders candidates them to manage the financial profile of the cluster as a whole. According to this, "corporate finance" is to become "something else" managing both the external financial needs of the cluster, along with the internal allocation of the financial resources. Indeed, the corporation is no more the institution to refer to, in order to understand the economics of transactions, while a "scaffolding approach" as suggested by [3], is preferable. Talking about "corporate-else finance" is then to be preferred to the classic corporate finance. This requires identifying the subject of the managerial finance activity. In this paper, we investigate the effective capability of the firm scaffolding, to appoint leaders in finance and to delegate them to the management of transactions, which are required to fund the entire cluster. To do this, we present a methodology to discriminate between the "leader" of a productive chain and the rest of the companies in this network (the suppliers). The idea is based on the discovery by [5], who show that in the Italian North East District of Treviso there appears to be a clear "Production Chain Network" where the leading firm (LF) finances eight supplier firms (SF). The inner aim of the paper is to discover if capitals prefer to flow to companies as legal entities (i.e. corporate finance),or to clusters as real entities (corporate-else finance). We think it is important to delve into this puzzle because, if it exists, banks are better off determining the merit of credit based on the "overall" network and not based on single entities in the network.

## 2 Theoretical Model and Robustness Checks

Specifically, in order to identify the leading firm (LF) and the supplier firms (SF) in the network, we hypothesize that the LF finances the rest of the network (its suppliers). To the contrary, we hypothesize that the SA does not finance the rest of the network. Figures 1 and 2 show the differences in the capital structure of both

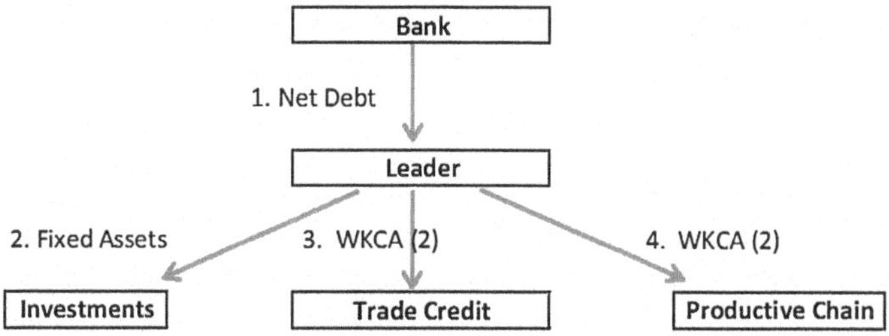


**Fig. 1** Leading firm (LF) in a productive chain

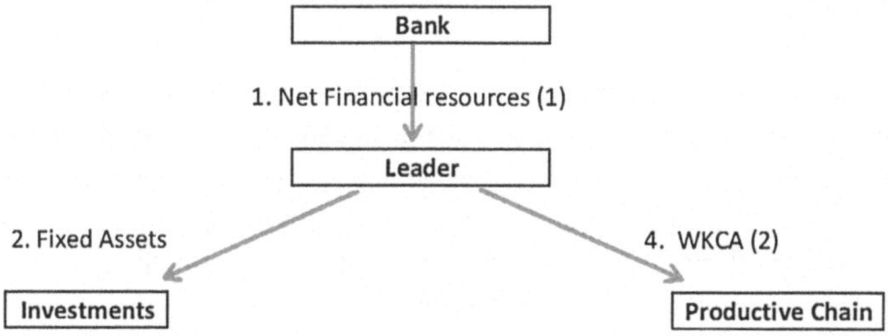


**Fig. 2** Standing Alone firm (SA)

LF and SA. LF presents additional working capital compared to SA because part of the capital raised from banks (net financial resources) is transferred to its suppliers, which are part of the productive chain network.

$$Net\ Financial\ Resources = Equity + Net\ Debt \tag{1}$$

$$WKCA = Working\ Capital = Debtors + Inventory - Creditors \tag{2}$$

Based on a number of theoretical simulations on the firms balance sheet, we are able to develop the following hypothesis: 1. LF firms show a positive correlation between absolute working capital intensity and absolute weight of financial resources ($\rho_1$); 2. LF firms show a positive correlation between relative working capital intensity and absolute weight of financial resources ($\rho_2$); 3. SA working capital is greater than that of LF; 4. SA Net financial resources are greater than those of SF; 5. There is a decreasing working capital intensity of SA. We test the above hypothesis on a sample of 13,391 firms incorporated in three regions of North East Italy (Veneto, Friuli Venezia Giulia and Trentino Alto-Adige) and with balance sheet data for every year from 2006 to 2012. We are able to identify 553 LF, 1.115 SA and 3.334SF. In order to check the validity of the theoretical model, we perform panel regressions which analyze the relation between the return of investment and a series of variables which try to capture the risk level of firms in the analysis [4] over the period from 2007 to 2012. The panel regression analysis is performed on four different samples: the total population, the manufacturing sector, the standing alone-SA cluster and the supplier firms-SF cluster and the leader firms-LF cluster. The results show that both the sub-samples identified by the SA and LF clusters present a higher Adjusted R-squared estimation, moving from 0.15 (manufacturing firms sub sample) to 0.73 in the SA cluster and 0.65 in the LF cluster when including the autoregressive component ($ROI_{t-1}$). Similarly, the estimation parameter moves from 0.06 to 0.64 in SA and 0.55 in LF when the autoregressive component ($ROI_{t-1}$) is excluded. Additionally, as expected, the Hannan-Quinn criterion for the above two clusters decreases in value. Thus, we conclude that results of the panel regressions analysis seem to confirm the soundness of the Productive Chain Networks identification method.

## 3 Conclusions

Even if clusters can be evidenced only by concrete case studies, this paper tries to investigate whether the hypothesis of corporate-else finance can be supported by empirical evidence. Leaders are identified recurring to a methodology based on the hypothesis of polarized farming of productive farming inside the clusters. This generates specific capital intensity levels both for working capital and for fixed capital; the relationship existing between their relative intensity and the capital flows can be proof of a leadership even in financial functions. Testing the methodology in the very dense area, in terms of number of firms of the North-East Italy, seems to detect correctly the leaders (LF). Financial reports depict their specificities, if compared with the other cluster firms and the stand alone ones. Evidences demonstrate that LF receive more debt resources than any others, and that they pay less for these resources. This results in a substitution of the banks role in the selection of investments. Hence, banks that are able to explore firms boundaries among networks, can surely improve their ability to estimate the merit of credit of such firms in two ways: (1) by improving the efficiency of rating models and (2) by considering both possible network re-allocation of financial resources and economic interdependencies.

## References

1. Barney, J.B., Ouchi, W.G.: Organizational Economics. Jossey-Bass, San Francisco (1986)
2. Coase, R.H.: The nature of the firm. Econ., N. S. **4**(16), 386–405 (1937)
3. Gurisatti, P.: MBA Lecture. @FAU in UCF-Venice (2013)
4. Mantovani, M.G., Daniotti, E.: Valori e Capitali per un Nuovo Patto di Sviluppo del Sistema—il Caso Treviso. Edizioni Ca' Foscari (2012)
5. Mantovani, G.M., Mestroni, M., Basilico, E.: Which is worth more for the merit of credit? What a company did or what it will do in the future? Evidence from the credit system in the North Eastern Italian District. Working Paper (2013)
6. Nkongolo-Bakenda, J.-M.: Inter-firm networking propensity in small and medium-sized enterprises (SMEs). J. Entrep. Finance **7**(1), 99–119 (2002)

# BEKK Element-by-Element Estimation of a Volatility Matrix. A Portfolio Simulation

**Alessia Naccarato and Andrea Pierini**

**Abstract** The use of a BEKK (Baba-Engle-Kraft-Kroner) model is proposed to estimate the volatility of a set of financial historical series with a view to the selection of a stock portfolio. An individual element on the diagonal of the volatility matrix is estimated by applying the model to the series of log returns both of the share $i$ to which it refers and of the market index. An extra-diagonal element is instead estimated by using in the model the covariances between the series of log returns of the two shares $i$ and $j$ to which the element of the volatility matrix corresponds.

The procedure proposed for the estimation of volatility was applied to the series of monthly stock log returns of 150 shares of major value traded on the Italian market between 1 January 1975 and 31 August 2011 and the Markowitz portfolio is simulated.

**Keywords** BEKK model · CVAR model · Markowitz portfolio · Simulation

## 1 Introduction

The problem that arises in the selection of a stock portfolio generally regards estimating the volatility of log returns, which means estimating a variance-covariance matrix [1]. The problems to be addressed in defining a portfolio are, however, of a different nature. First it is necessary to take the entire set of shares and identify the subset from which those to be included in the portfolio are selected on the basis of ranking. Then the volatility matrix, and hence the investment risk, is estimated. Finally, the portfolio is defined according to the information obtained in the first two phases. The first problem is addressed in this work by proposing a selection criterion based on the differences between the log intrinsic value of the shares and the log returns. The volatility matrix of the shares selected is then estimated through

A. Naccarato · A. Pierini (✉)
Department of Economics, University of Roma Tre, via S. D'Amico 77, 00145 Rome, Italy
e-mail: andrea.pierini@uniroma3.it

A. Naccarato
e-mail: alessia.naccarato@uniroma3.it

C. Perna, M. Sibillo (eds.), *Mathematical and Statistical Methods for Actuarial Sciences and Finance*, DOI 10.1007/978-3-319-05014-0_34,

combined use of the Cointegrated Vector Autoregressive (CVAR) model [2, 5, 7] and the Baba-Engle-Kraft-Kroner (BEKK) model [3]. Once the volatility matrix has been estimated by solving a problem of quadratic optimization, it is possible to establish the proportions in which each of the shares selected is to be purchased, i.e. to select a portfolio.

## 2 Selection of Shares and Estimation of the Volatility Matrix

The need to select a (possibly large) subset of $n$ shares for inclusion in the portfolio on the basis of their log returns stems from the fact that the entire set of shares available on the market is so large that it would be impossible to define a portfolio a through simultaneous study of all their log returns. It therefore makes sense to concentrate on the subset of the "best" shares, defined here as those with the greatest difference between log intrinsic value and log returns. The number $n$ of shares to be selected is established through application of a criterion (efficient frontier) that identifies the combination of shares offering the returns with minimum risk for every fixed $n$ but with the composition varying in terms of the quantity of each share to be purchased. The first problem to be addressed in the selection of shares is therefore the calculation of the difference between the log intrinsic value of the share and its log return. It should be pointed out, however, that this is in any case a problem of prediction, as the point of interest is the future value of the difference between the two magnitudes considered. This makes it possible to decide which share to select at the moment of investment. It is therefore necessary to estimate a model that makes this prediction possible. A CVAR(p) model is adopted in order to estimate both the log intrinsic value of the share and its log return. The starting point is the $K = 150$ series, regarding the log returns $R_{k,t}$ on the shares, and the average log return of the market $R_{M,t}$, $t = t_k, \ldots, T$, $k = 1, \ldots, K$. For each series, the CVAR(p) model is considered for the random vector $y_t = [y_{1,t}, y_{2,t}]' = [R_{k,t}, R_{M,t}]'$ as

$$\Delta y_t = \eta_t + \Pi y_{t-1} + A_1 \Delta y_{t-1} + \cdots + A_{p-1} \Delta y_{t-p+1} + u_t \qquad (1)$$

where $\Delta$ indicates the usual difference operator, $\eta_t = \eta_0 + \eta_1 \cdot t$, $A_i$ is the $2 \times 2$ matrix, $\Pi$ is the matrix of parameters containing information on the cointegration of the series [5], $i = 1, \ldots, p-1$, $\eta_0, \eta_1, A_i, \Pi$ are the unknown coefficients and $u_t = [u_{1,t}, u_{2,t}]'$ is the vector of errors such that $u_t \sim N(0, \Sigma_u)$. It should be noted that the CVAR(p) model is chosen to estimate the unknown coefficients of Eq. (1) because it makes it possible to consider the possible presence of integration or cointegration between the two components of the random vector $y_t$. Note also that when the series present neither cointegration nor integration, Eq. (1) is not informative and it becomes necessary to estimate a Vector Autoregressive (VAR) model. The same procedure is also adopted to estimate the model for log intrinsic value, where use is made not only of the historical series of log intrinsic value for the period under examination, but also of the average log intrinsic value of the sector of economic activity to which the share belongs. Once the two magnitudes have been estimated for every share $i$, all the differences that present positive values of differences and log return at the same time are also estimated. With the number $n$ of shares to be selected set initially at 10, the volatility matrix is estimated element by element. In particular, a two-step procedure is adopted to estimate the variance of the log return of share $i$ (element $i$ on the main diagonal of the matrix). First, a CVAR(p) model is estimated in which the historical series considered are the log return of the share $i$ itself and the log return of the market index [7]. Second, if the ARCH test [5] carried out on the residuals of the CVAR model estimated in step 1 indicates the presence of heteroscedasticity, a BEKK model is applied to the same residuals, which makes it possible to interpret the temporal dynamics of the variances of the log return of share $i$ [3]. In order to estimate the extra-diagonal elements of the volatility matrix (covariances between the log returns of two shares $i$ and $j$), the same procedure is used with the difference that the CVAR model is applied to the series of the log returns of the two shares. Here too, if the ARCH test indicates the presence of heteroscedasticity, a BEKK model is estimated. It should be noted that it is possible to demonstrate that the two-step procedure converges asymptotically on the simultaneous estimation of the elements of the matrix [5]. Moreover, since the estimation of the entire matrix of volatility is obtained asymptotically as a composition of consistent and increasingly efficient estimates [5], it presents the same characteristics. The size $n$ of the portfolio is increased by means of an iterative procedure until all the "best" shares are included in the portfolio.

## 3 Results and Conclusions

The model put forward was applied to the 150 shares of highest value on the Italian stock market. The maximum number of shares in the portfolio proved to be 25. Figure 1 top left shows the volatility and the log return obtained by solving the Markowitz optimization problem for variation of the expected return $R_{p,T+1}$ and the dimension $n$ of the portfolio ($n = 10, 11, \ldots, n_{max}$). The portfolio risk tends to decrease as $n$ increases [6]. Figure 1 top right shows the efficient frontiers (defined by the part of every curve continuing upward from X) obtained by solving the Markowitz optimization problem for variation of the expected return $R_{p,T+1}$ and the dimension $n$ of the portfolio ($n = 10, 11, \ldots, n_{max}$). The optimal risk from a risk-averse standpoint corresponds to $n = 25$, i.e. to the point on the curve furthest to the left in Fig. 1 to right, indicated with the symbol F. The portfolio thus identified presents an average monthly return of 0.00993, a standard monthly deviation of 0.0630 and a Sharpe index value of 0.15771. Figure 1 bottom presents the estimates of the elements of the volatility matrix and shows that the risk is mostly due to the variances of the shares, to which the highest peaks correspond. It is evident, however, that the values of variance and covariance are comparable for some subsets of shares. This suggests that it could prove useful, in order to reduce computational complexity, to take covariance into consideration only for specific subgroups of shares and variance alone for the others. It therefore becomes necessary

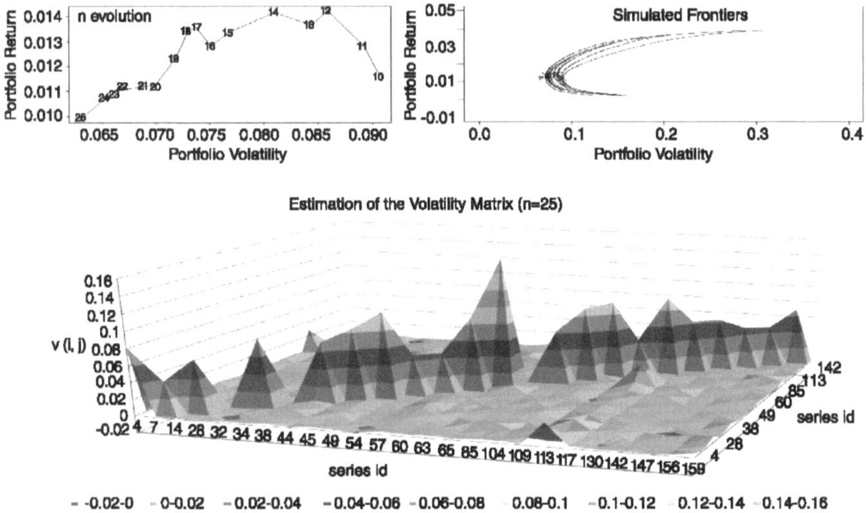


**Fig. 1** *Top Left*: *n* evolution. *Top Right*: Portfolio frontiers simulation. *Bottom*: Volatility estimation

to develop a criterion, based for example on the Granger principle of causality or on analysis of cross-correlation [4], in order to identify the groups of shares to be addressed in a different way.

## References

1. Brown, K., Reily, F.: Investment Analysis and Portfolio Management. South-Western College Publishing (2008)
2. Campbell, J.Y., Chan, Y.L., Viceira, L.M.: A multivariate model of strategic asset allocation. J. Financ. Econ. **67**(1), 41–80 (2003)
3. Engle, R.F., Kroner, K.F.: Multivariate simultaneous generalized ARCH. Econom. Theory **11**(1), 12–50 (1995)
4. Eichler, M.: Graphical modelling of multivariate time series. Probab. Theory Relat. Fields **153**, 233–268 (2012)
5. Lutkepohl, H.: New Introduction to Multiple Time Series Analysis. Springer, Berlin (2007)
6. Nyholm, K.: Strategic Asset Allocation in Fixed Income Markets. Wiley, New York (2008)
7. Tsay, R.S.: Analysis of Financial Time Series. Wiley, New York (2005)

# The Effects of Curvature and Elevation of the Probability Weighting Function on Options Prices

**Martina Nardon and Paolo Pianca**

**Abstract** We evaluate European financial options under continuous cumulative prospect theory. Within this framework, it is possible to model investors' attitude toward risk, which may be one of the possible causes of pricing errors. We focus on probability risk attitudes and use alternative probability weighting functions. In particular, curvature of the weighting function models optimism and pessimism when one moves from extreme probabilities, whereas elevation can be interpreted as a measure of relative optimism. The *constant relative sensitivity* weighting function is the only one, amongst those in the literature, which is able to model separately curvature and elevation. We are interested in studying the effects of both these features on options prices.

**Keywords** Cumulative prospect theory · Curvature · Elevation · European option pricing

## 1 Introduction

*Prospect theory* (PT) has recently begun to attract attention in the literature on financial options valuation; when applied to option pricing in its continuous cumulative version, it seems a promising alternative to other models, for its potential to explain option mispricing with respect to Black and Scholes [2] model.

According to prospect theory, individuals do not always take their decisions consistently with the maximization of expected utility. Decision makers are risk averse when considering gains and risk-seeking with respect to losses. They are loss averse: people are much more sensitive to losses than they are to gains of comparable magnitude. Gambles are evaluated considering potential gains and losses relative to a *reference point*, rather than in terms of final wealth. Individuals have also biased

M. Nardon (✉) · P. Pianca
Department of Economics, Center for Quantitative Economics, Ca' Foscari University of Venice, Cannaregio 873, 30121 Venice, Italy
e-mail: mnardon@unive.it

P. Pianca
e-mail: pianca@unive.it

C. Perna, M. Sibillo (eds.), *Mathematical and Statistical Methods for Actuarial Sciences and Finance*, DOI 10.1007/978-3-319-05014-0_35, 

probability estimates; they tend to underweight high probabilities and overweight low probabilities.

PT in its formulation proposed by [4] is based on the subjective evaluation of *prospects*. Prospects assign to any possible outcome a probability; originally PT deals only with a limited set of prospects. Risk attitude, loss aversion and subjective probabilities are described by two functions: a value function $v$ and a weighting function $w$, which models probability perception.

*Cumulative prospect theory* (CPT) developed by [6] overcomes some drawbacks (such as violation of stochastic dominance) of the original PT. A value function alone is not able to capture the full complexity of observed behaviors: the degree of risk aversion or risk seeking appears to depend not only on the value of the outcomes but also on the probability and ranking of outcome. Let $x$ denote an outcome, subjective values $v(x)$ are not multiplied by objective probabilities, but using *decision weights*. Decision weights are differences in transformed (through a weighting function) cumulative probabilities of gains or losses.

The shape of the value function and the weighting function becomes significant in describing actual choice patterns. It is also relevant to separate gains from losses, as negative and positive outcomes may be evaluated differently: the function $v$ is typically convex in the range of losses and concave and steeper in the range of gains, whereas subjective probabilities may be evaluated through a weighting function $w^-$ for losses and $w^+$ for gains.

Specific parametric forms have been suggested for the value function. A function which is used in many empirical studies is

$$\begin{aligned} v^- &= -\lambda(-x)^b \quad && x < 0, \\ v^+ &= x^a && x \geq 0, \end{aligned} \tag{1}$$

with positive parameters which control risk attitude ($0 < a \leq 1$ and $0 < b \leq 1$) and loss aversion ($\lambda \geq 1$); $v^-$ and $v^+$ denote the value function for losses and gains, respectively.

In financial applications, and in particular when dealing with options, prospects may involve a continuum of values; hence, prospect theory cannot be applied directly in its original or cumulative versions. Davies and Satchell [3] provide the continuous cumulative prospect value:

$$V = \int_{-\infty}^{0} \Psi^-\big[F(x)\big] f(x) v^-(x)\, dx + \int_{0}^{+\infty} \Psi^+\big[1 - F(x)\big] f(x) v^+(x)\, dx, \tag{2}$$

where $\Psi = \frac{dw(p)}{dp}$ is the derivative of the weighting function $w$ with respect to the probability variable, $F$ is the cumulative distribution function (cdf) and $f$ is the probability density function (pdf) of the outcomes.

## 2 The Weighting Function

Prospect theory involves a probability weighting function which models probabilistic risk behavior. A weighting function $w$ is uniquely determined, it maps the probability interval [0, 1] into [0, 1], and is strictly increasing, with $w(0) = 0$ and $w(1) = 1$. The *curvature* of the weighting function is related to the risk attitude towards probabilities. Empirical evidence suggests a particular shape of probability weighting functions: small probabilities are overweighted $w(p) > p$, whereas individuals tend to underestimate large probabilities $w(p) < p$. This turns out in a typical *inverse-S shaped* weighting function: the function is initially concave (probabilistic risk seeking or *optimism*) for probabilities in the interval $(0, p^*)$, and convex (probabilistic risk aversion or *pessimism*) in the interval $(p^*, 1)$, for a certain value of $p^*$. A linear weighting function describes probabilistic risk neutrality or objective sensitivity towards probabilities, which characterizes Expected Utility. Empirical findings indicate that the intersection (*elevation*) between the weighting function and the 45 degrees line, $w(p) = p$, is for $p^*$ in the interval $(0.3, 0.4)$.

The *constant relative sensitivity* (CRS) weighting function proposed by [1]

$$w(p) = \begin{cases} \delta^{1-\gamma} p^{\gamma} & 0 \le p \le \delta \\ 1 - (1-\delta)^{1-\gamma}(1-p)^{\gamma} & \delta < p \le 1, \end{cases} \tag{3}$$

is the only one, amongst those in the literature, which is able to capture separately the effect of curvature and elevation: the parameter $\delta$ controls elevation and may be interpreted as an index of relative optimism, whereas $\gamma$ measures relative sensitivity of the weighting function.[1]

## 3 European Options Valuation

We evaluate European financial options within continuous CPT, under the hypothesis that the underlying price dynamics is driven by a geometric Brownian motion. Versluis et al. [7] provide the prospect value of writing call options, considering different time aggregation of the results. Their results are extended to the case of put options in [5]; the authors consider the problem both from the writer's and holder's perspective.

In this contribution, we perform a wide sensitivity analysis on call and put options values, applying alternative weighting functions. In particular, when applying the weighting function (3), we let vary the parameters $\gamma \in [0.7, 1.0]$ and $\delta \in [0.3, 0.4]$, considering also different sensitivity to probability risk for positive and negative outcomes. For the value function, we compared different parameters sets, ranging from TK sentiment (see [6]) to *moderate sentiment*; a linear function (with $a = b = 1$ and $\lambda = 1$) is considered as a limiting case (no sentiment). We computed the option prices for several values of the volatility and the strike price $X$. Numerical results suggest that option prices are increasing with $\delta$ (elevation) within the interval $[0.3, 0.4]$, whereas the effect of $\gamma$ (curvature) is non-trivial, depending on the

[1]The index of relative sensitivity of function (3) is constant on the interval $(0, 1)$ and is equal to $1 - \gamma$.

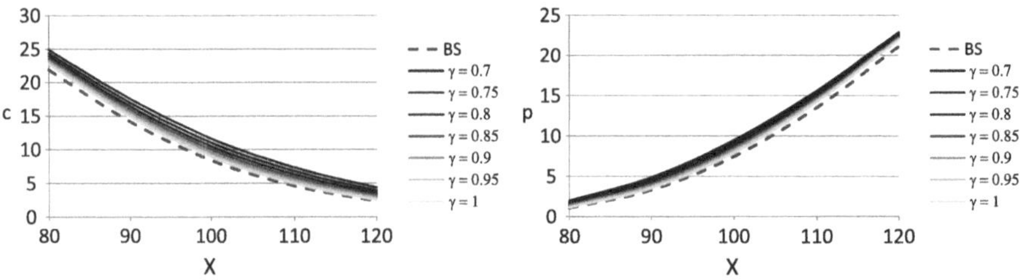


**Fig. 1** Sensitivity of the call (*left*) and put (*right*) option prices (writer's position in the time-aggregated model) to the curvature of the probability weighting function, $\gamma \in [0.7, 1.0]$, with $\delta = 0.325$. BS is the Black-Scholes price (with $\gamma = 1$, $a = b = 1$, and $\lambda = 1$). The option parameters are: $S_0 = 100$, $X \in [80, 120]$, $r = 0.01$, $\sigma = 0.2$, $T = 1$; the parameters of the value function are: $a = b = 0.976$, and $\lambda = 1.125$

moneyness and the model (time-aggregated or segregated) which is used. As an example, Fig. 1 shows some results for the call and put options in the time-aggregated model; in these cases, option premia are decreasing with curvature. Detailed results are reported and described in a separate paper.

## References

1. Abdellaoui, M., L'Haridon, O., Zank, H.: Separating curvature and elevation: a parametric probability weighting function. J. Risk Uncertain. **41**, 39–65 (2010)
2. Black, F., Scholes, M.: The pricing of options and corporate liabilities. J. Polit. Econ. **81**(3), 637–654 (1973)
3. Davies, G.B., Satchell, S.E.: The behavioural components of risk aversion. J. Math. Psychol. **51**(1), 1–13 (2007)
4. Kahneman, D., Tversky, A.: Prospect theory: an analysis of decision under risk. Econometrica **47**(2), 263–292 (1979)
5. Nardon, M., Pianca, P.: A behavioural approach to the pricing of European options. In: Corazza, M., Pizzi, C. (eds.) Mathematical and Statistical Methods for Actuarial Sciences and Finance, pp. 217–228. Springer, Milano (2013)
6. Tversky, A., Kahneman, D.: Advances in prospect theory: cumulative representation of the uncertainty. J. Risk Uncertain. **5**, 297–323 (1992)
7. Versluis, C., Lehnert, T., Wolff, C.C.P.: A cumulative prospect theory approach to option pricing. Working paper. LSF Research Working Paper Series 09–03, Luxembourg School of Finance (2010)

# A Multivariate Approach to Project the Long Run Relationship Between Mortality Indices for Canadian Provinces

**Achille Ntamjokouen, Steven Haberman, and Giorgio Consigli**

**Abstract** The cointegration approach is proposed to model cross-province mortality indices within Canada. We apply and compare the vector autoregressive model (VAR) and the vector of error correction model (VECM) derived from cointegrated models for males and females. Relying on the Johansen cointegration test, the analysis shows clearly that there is a dependence among provincial mortality indices. The two models fit well the females data. However, poor performance has been revealed for men beyond 10 years horizons. We project the mortality indices from both models and compute the annuity from the forecasts. We project the mortality indices from both models and compute the annuity from the forecasts.

**Keywords** Mortality indices · VAR · VECM · Pricing by cohorts

## 1 Introduction

Over the last century, life expectancy has been increasing with declining mortality rates as reported by [6] for G7 countries. The Lee Carter model [1] which is an extrapolation method is used to quantify this phenomenon. It is not only a measure of an age period model but it measures also trends on the evolution of mortality by age. It presents drawbacks such as the uncertainty deriving from errors in the quantification of pattern of deviation coefficients. Further, a higher factor term could be incorporated into the model. A growing literature, based on the cointegration approach, which tackles these deficiencies, is being explored by articles such as [3, 5].

A. Ntamjokouen (✉) · G. Consigli
University of Bergamo, via Salvecchio 19, 24129 Bergamo, Italy
e-mail: achille.ntamjokouen@unibg.it

G. Consigli
e-mail: giorgio.consigli@unibg.it

S. Haberman
Cass Business School, City University London, 106 Bunhill Row, EC1Y 8TZ London, UK
e-mail: s.haberman@city.ac.uk

C. Perna, M. Sibillo (eds.), *Mathematical and Statistical Methods for Actuarial Sciences and Finance*, DOI 10.1007/978-3-319-05014-0_36,

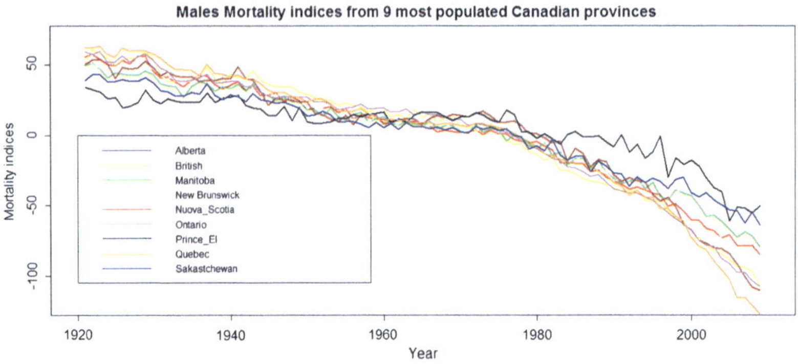


**Fig. 1** Males mortality indices from each of the 9 provinces in Canada

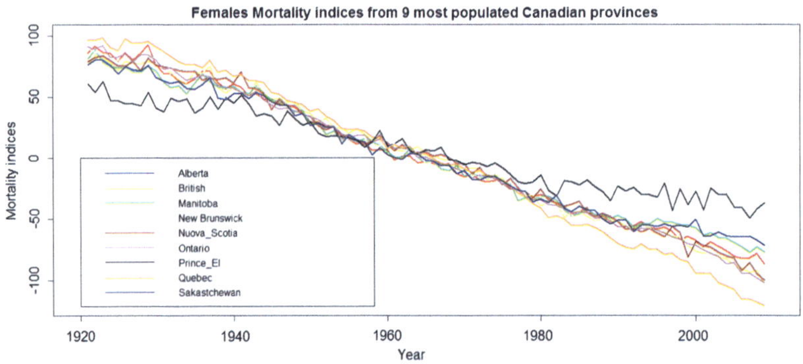


**Fig. 2** Females mortality indices from each of the 9 provinces in Canada

Issues arise regarding long term relationship among the 12 provinces in Canada. Our goal is to explore evidence of common trends in mortality indices in the long term. Given that cointegration on mortality indices has been applied only on two populations for UK (see [7]), we want to extend it to more than two. From the Lee Carter model, we retrieve the mortality indices from each singular provincial mortality and then apply the cointegration approach. Mortality data are provided by Canadian Human Mortality Database which the website is www.bdlc.umontreal.ca/chmd.

We use mortality data from the 9 most populated provinces of Canada which include: Alberta, British Columbia, Manitoba, Nova Scotia, News Brunswick, Prince Edward Island, Ontario, Quebec, Sakastchewan. The provincial mortality indices show a decreasing trend for all the provinces from 1921 to 2009 (see Figs. 1 and 2). Further observations show that trends are amplificated from 1970 onwards.

The methodology includes 6 main points. The first regards the determination of the order of integration for each of the variable. The second illustrates the computation of the optimal value of lag of the VAR while the third points out estimations from various models. The fourth is devoted to the Johansen cointegration test. The fifth regards the backtesting of both VECM and the VAR models. The sixth deals with the forecasting of derived models. And finally the models will be used to calculate Actuarial pricing value for group of cohorts.

## 2 Methods and Results

The estimation starts by submitting the mortality indices to the unit root test. The series show clearly that variables are integrated of order 1 under the 3 main unit root test including ADF, PP as well as KPSS.

Following [4], We observe that AIC indicates 6 lags for both females and males whereas other information criterion such as HQ, FPE and SC indicate only 1. Following [2], the optimal lag length is 1 as suggested by SC criteria for both cases.

The vector of autoregression for p lags is written in [2] as:

$$k_t = A_0 + A_1 k_{t-1} + A_2 k_{t-2} + \cdots + A_p k_{t-p} + e_t \tag{1}$$

where $k_t = (k_{1t}, k_{2t}, \ldots, k_{Kt})$ for $k = 1, \ldots, K$ time series, $(A_0 \ldots A_i)$ are the coefficients and $e_t$ is white noise. The long run specification of VECM is in the form:

$$\Delta k_t = \Gamma_1 \Delta k_{t-1} + \Gamma_2 \Delta k_{t-2} + \cdots + \Gamma_{p-1} \Delta k_{t-p+1} + A_0 + e_t \tag{2}$$

where $\Gamma_i = -(I - A_1 - \cdots - Ai), i = 1, \ldots, (p-1), \Pi = -(I - A_i, - \cdots - A_p)$ is a $N$-dimensional time series, $A_0$ is the intercept term, $e_t$ is white noise. We run the estimations for VAR and VECM models with lag = 1 for all provinces. We experiment the Johansen methodology for trace and Eigenvalue tests. Females mortality indices identify 5 cointegrated equations for both tests. Similar analysis for males mortality indices show respectively 3 and 4 for respectively trace and eigenvalues. It can be deduced that there are 4 common stochastic trends in the case of females. Males analysis rather indicate respectively 6 and 5 common factors. They reveal a clear dependence between mortality indices in a country. In the following part of this document, we focus only on the results obtained from the trace test.

We carry out diagnostics tests on residuals. In the case of VAR, the test of non autocorrelation(Portmanteau test) shows respectively for female and male p-value as 0.68 and 0.81. Additionally the test of normality shows p-value 0.31 for females and 0.18 for males. In the case of VECM, the test of autocorrelation(Portmanteau test) shows respectively for female and male p-value as 0.75 and 0.97. Also, the test of normality shows p-value as 0.16 for females and 0.04 for males. We remark from these tests that there is evidence of normality and non autocorrelation of the residuals for the two models. The VAR for the two genders group and VECM for females appear to be well behaved with white noise disturbances. However the normality test for VECM is rejected in the case of males. A higher order lag may provide better results with respect to this test. But, Since this study is interested in the forecasting performance of the model, fewer parameters are expected to be included.

In order to evaluate the performance of these models out-of-sample, We have backtested the models VAR and VECM by using out-of-samples. The results from MAPE(Mean absolute of percentage of error) are described in Table 1.

**Table 1** The average MAPE for models VAR and VECM for the 9 provinces: The first two columns refer to females and the last two to males

| Out-of-sample | VAR(Females) | VECM(Females) | VAR | VECM |
|---|---|---|---|---|
| $h = 2005–2009$ | 5.63 % | 5.13 % | 6.85 % | 5.73 % |
| $h = 2002–2009$ | 6.63 % | 6.52 % | 9.47 % | 10.96 % |
| $h = 2000–2009$ | 12.89 % | 7.43 % | 8.42 % | 22.91 % |
| $h = 1995–2009$ | 16.38 % | 9.79 % | 10.66 % | 2.45 % |
| $h = 1990–2009$ | 19.36 % | 15.14 % | 29.67 % | 24 % |
| $h = 1984–2009$ | 21.77 % | 16.80 % | 39.80 % | 30.01 % |

1. We observe from Table 1 that VECM performs better than VAR for females. Also, the backtesting for females from both models presents good performance accuracy overall.
2. It is uncertain for the first 3 periods as to males for first 3 samples which model is better. Furthermore beyond the 10 years period, errors are too large. This is due to the fact that they cope volatility of future mortality indices only partially.

We project mortality indices from the 9 provinces in 50 years ahead. We observe that the forecasts 95 % confidence interval of mortality indices for all the provinces in Canada are narrow with VAR model (see Figs. 3 and 4). This is due to the fact that the confidence interval does not allow for more quantification of mortality improvements. However, the predictions interval confidence from VECM is wider for both sexes (see Figs. 5 and 6). The VECM improves VAR and also the risk quantification is improved.

The computations of the price of annuities and life expectancy for cohorts 1960, 1970, 1980 are greater when we apply VECM rather than VAR and ARIMA.

## 3 Conclusion

In this paper, we investigate the dependence of mortality indices between the 9 most populated Canadian provinces through cointegration approach. Mortality multivariate times series of indices have been retrieved from Lee Carter model for both genders. Specifically, the Johansen test has been used to show dependence between provinces. We first compute the VAR then VECM model that show clearly a dependence between mortality indices. This econometric analysis allows to capture common trends in mortality across provinces. The two models applied to mortality indices work better for females than for males out-of-sample. They take into account just part of mortality improvements in the long run for males. The two models have been used then to project future mortality indices for Canadian provinces. The VECM allows for more quantification of mortality risk than VAR. Overall, the model is useful for modeling relationship between provincial mortality indices especially female gender.

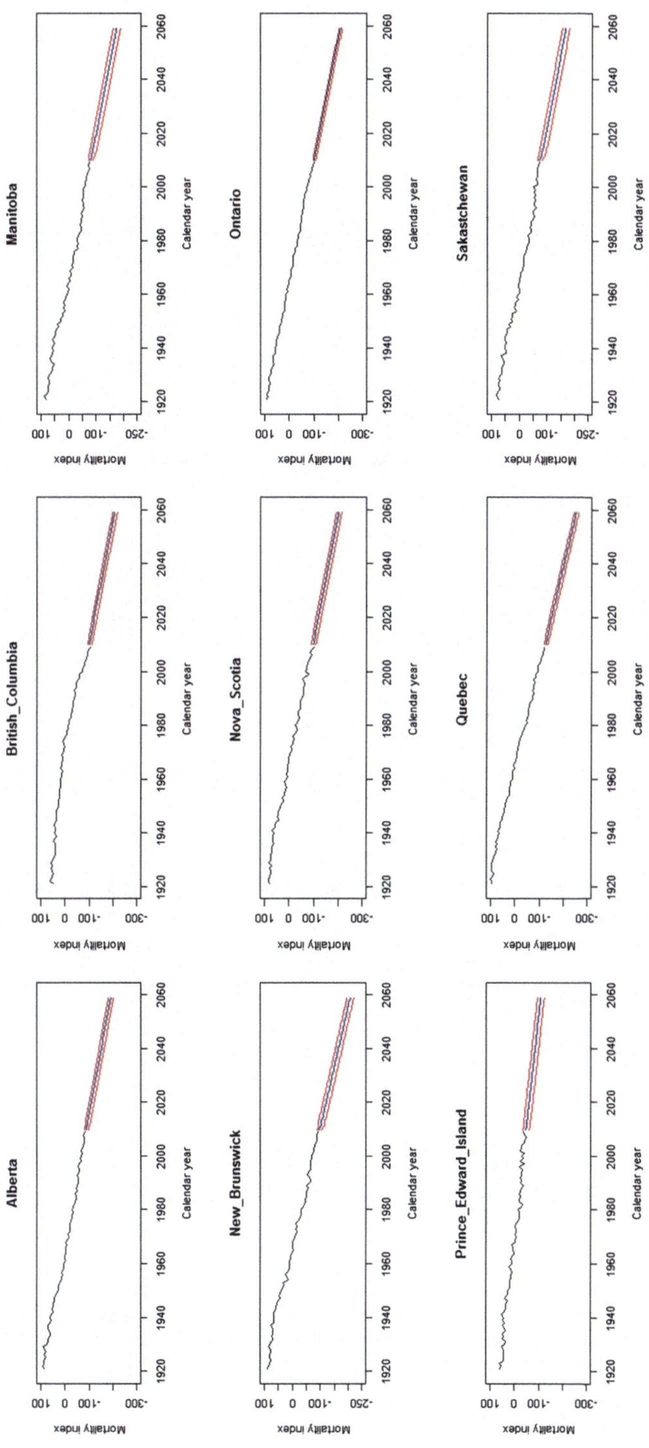


**Fig. 3** Projecting females mortality indices for all provinces with VAR models

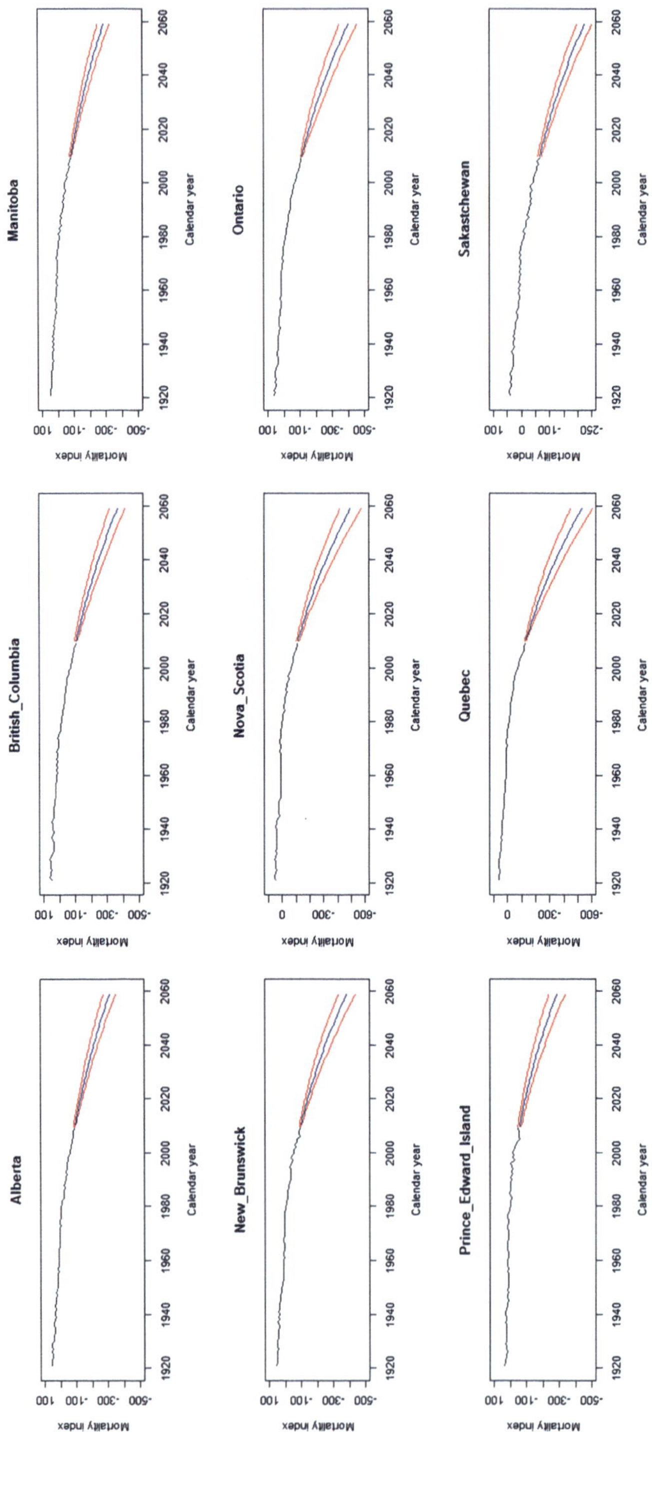


**Fig. 4** Projecting Males mortality indexes for all provinces with VAR models

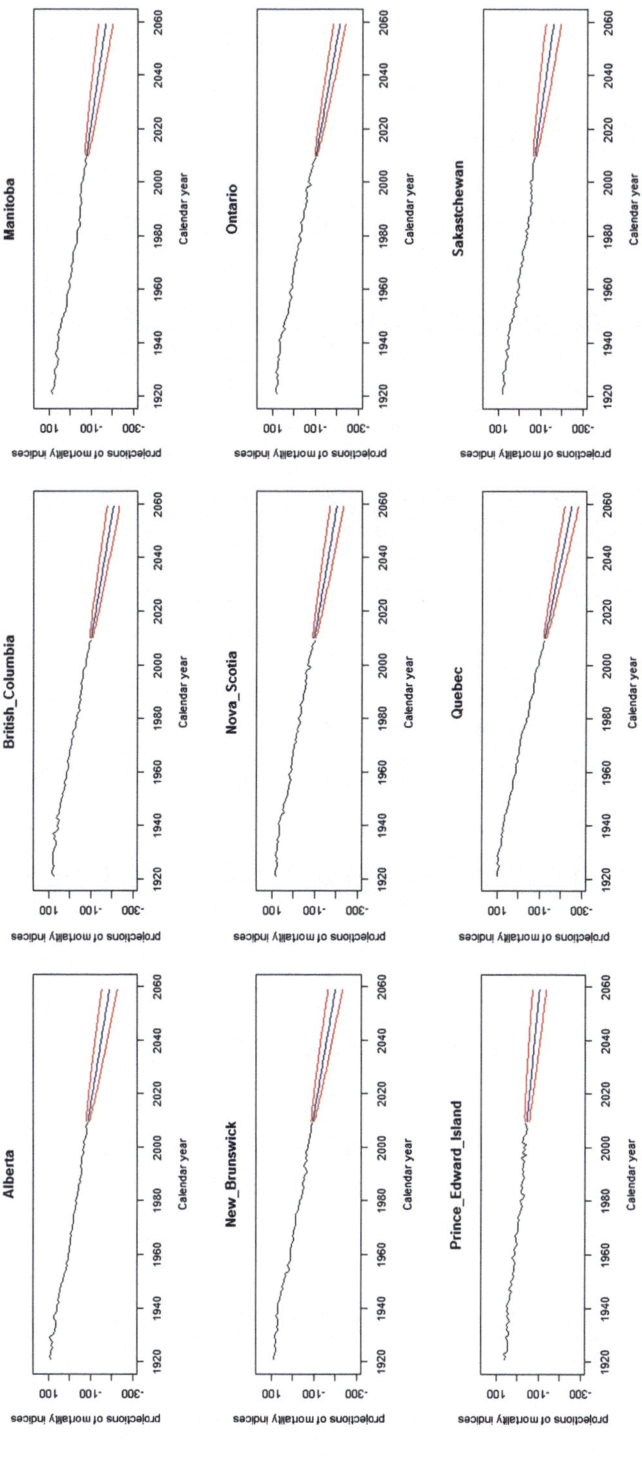


**Fig. 5** Forecasting Canadian females Mortality indexes from the Vector of Error Correction model

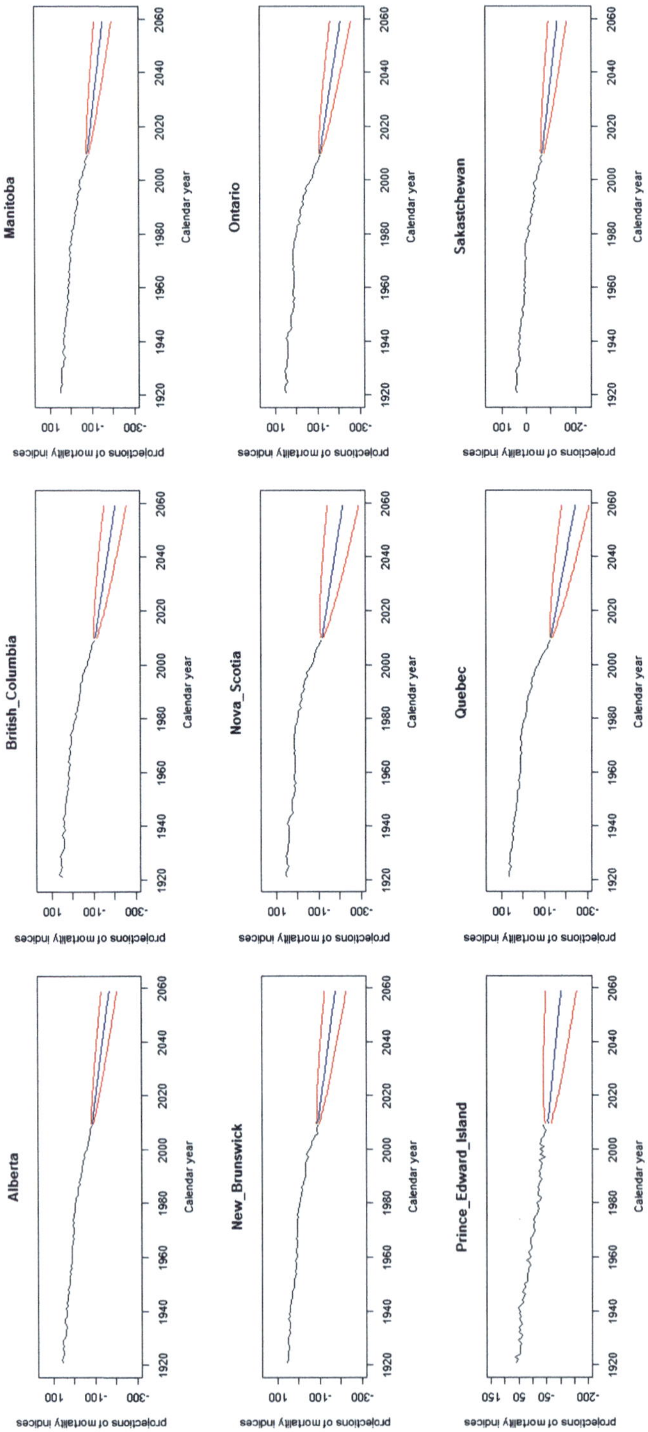


**Fig. 6** Forecasting Canadian Males Mortality indexes from the Vector of Error Correction model

## References

1. Lee, R.D., Carter, L.R.: Modeling and forecasting U.S. mortality. J. Am. Stat. Assoc. **87**, 659–675 (1992)
2. Lutkepohl, H.: New Introduction to Multiple Time Series Analysis. Springer, Berlin (2005)
3. Njenga, C., Sherris, M.: Longevity risk and the econometric analysis of mortality trends and volatility. Asia-Pac. J. Risk Insur. **5**, 1–52 (2011)
4. Pfaff, B.: VAR, SVAR and SVEC Models: implementation with the R package vars. J. Stat. Softw. **27**(4) (2008)
5. Salhi, Y., Loisel, S.: Joint modeling of portfolio experienced and national mortality: a co-integration based approach. Working Paper. Available from http://isfa.univ-lyon1.fr/stephane.loisel/ (2010)
6. Tuljapurkar, S., Puleston, C.O., Gurven, M.D.: Why men matter: mating patterns drive evolution of human lifespan. PLoS ONE **2**(8), e785 (2007). doi:10.1371/journal.pone.0000785
7. Zhou, R., Wang, Y., Kaufhold, K., Li, J.S.-H., Tan, K.S.: Modeling mortality of multiple populations with vector of error correction models: applications to solvency II. Working Paper (2012)

# Measuring and Managing the Longevity Risk: An Empirical Evidence From the Italian Pension Market

**Albina Orlando, Govanna di Lorenzo, and Massimiliano Politano**

**Abstract** This paper deals with the problem of quantifying the longevity impact for defined contribution pension funds in a stochastic environment. In the accumulation phase it is well known that, in presence of a benefit guarantee, the investment risk dominates the demographic one. However, if the generic subscriber life expectancy increases, it is very likely that, in the decumulation phase, the wealth accrued will not be able to cover the liabilities of the fund. For this reason, the fund will be forced to set aside more resources in order to front its liabilities exposing itself to greater financial risk. In this paper we study the interaction between financial risk and longevity: based on the Italian experience for both financial and demographic factors, this work aims to measure the impact of longevity on the financial factor.

**Keywords** Pension funds · Longevity risk · Forecasting mortality · Financial risk

## 1 Introduction

During the 20th century, human life expectancy have considerably increased for the population of many developed countries. Although past trends suggest that further changes in the level of mortality are to be expected, the future improvements of life expectancy are uncertain and difficult to be predicted. This uncertainty about the future development of mortality gives rise to longevity risk. The real challenge for pension systems consists precisely in the design of products able to absorb any adverse events concerning the future mortality. In other words, the challenge is how to

A. Orlando
Istituto per le Applicazioni del Calcolo Mauro Picone, Consiglio Nazionale delle Ricerche, Naples, Italy
e-mail: a.orlando@na.iac.cnr.it

G. di Lorenzo · M. Politano (✉)
Dipartimento di Scienze Economiche e Statistiche, University of Naples, Naples, Italy
e-mail: politano@unina.it

G. di Lorenzo
e-mail: giodilor@unina.it

C. Perna, M. Sibillo (eds.), *Mathematical and Statistical Methods for Actuarial Sciences and Finance*, DOI 10.1007/978-3-319-05014-0_37,

deal with the longevity risk. When we treat benefits depending on the survival of a certain number of individuals, the calculation of present values requires an appropriate projection of mortality in order to avoid an underestimation of future liabilities. Actually, this problem is deeply felt by private pension funds. Although the pension market is not well developed in the European countries, the reduction of the intervention field of public systems, due to the main goal of the cost containment and the gradual shifts from defined benefit schemes to defined contribution systems, suggest a growing interest of individuals for pension annuities. The main problem is to make the pension market attractive. For this reason, many pension fund providers focus in the issue of sharing the longevity risk between the annuitants and the annuity provider. In this field, many solution has been proposed: Denuit et al. [4] provides a solution trough securitization via longevity bonds, Denuit et al. [5] proposed a very interesting idea based on the reduction of annuity periodic payments, Pitacco [6] develops this concept and consider an adjustment factor to be applied to future benefits. In this paper, we look at the problem during the accumulation phase: that is, if future life expectancies of the pension participant increases, which means the plan is more likely to be underfunded, the fund should take more financial risk to accumulate sufficient funds to cover the claims. In other words, we can find a balance between the benefits reduction during the cumulation phase and the profitability increase during the accumulation phase. The paper is organised as follows: Sect. 2 introduces analytical developments. Section 3 defines the financial and mortality scenario. Finally, Sect. 4 offers numerical evidence.

## 2 The Model

Let us consider a defined contribution (DC) pension fund with an individual funding method, which pays off a capital amount resulting by a contribution accumulation process to the subscriber in presence of predecease, disability or old age. If we consider a benefit guarantee, the liability $b_h$ borne out by the fund in the year $h$ with respect to a generic subscriber is given by

$$b_h = \max\left\{W_h^A, W_h^{GAR}\right\} \tag{1}$$

where $W_h^A = \sum_{s=0}^{\xi} \frac{c_s}{a_s} a_\xi$ is equal to the share of the equivalent assets constituting the fund and $W_h^{GAR} = \sum_{s=0}^{\xi} n_s a_s (1+i)^{\xi-s}$ denotes the minimum guaranteed benefit. In this context, in case of death (d), disability (i) or old age (v), the liability of the fund toward a scheme member can be expressed as

$$W_0^L = \sum_{h=0}^{\xi} b_h \left({}_{h-1/1}q_x^{(d)} + {}_{h-1/1}q_x^{(i)}\right) e^{-\Delta(h)} + b_\xi \, {}_\xi p_x e^{-\Delta(\xi)} \tag{2}$$

where $\Delta(h) = \int_0^h r_u du$ is the accumulation function of the spot rate ${}_{h-1/1}q_x^{(d)}$, and ${}_{h-1/1}q_x^{(i)}$ respectively indicate the probability of death and disability in the time interval $[h-1, h]$, ${}_\xi p_x$ is the probability of reaching the pension age.

## 3 Financial and Mortality Scenarios

The valuation of the financial instruments involving the fund will be made assuming a two factor diffusion process obtained by joining Cox-Ingersoll-Ross (CIR) [2] model for the interest rate risk and a Black-Scholes (BS) [1] model for the stock market risk; the two source of uncertainty are correlated. The interest rate dynamics $\{r_t; t = 1, 2, \ldots\}$ is described by means of the diffusion process

$$dr_t = k^r(\theta - r_t)dt + \sigma^r \sqrt{r_t} dZ_t^r \tag{3}$$

where $k^r(\theta - r_t)$ is the drift of the process, $\sigma^r \sqrt{r_t}$ is the diffusion coefficient $Z_t^r$ is a Standard Brownian Motion. Clearly, it is very important the specification of the reference portfolio dynamics. The diffusion process for this dynamics is given by the stochastic differential equation

$$dS_t = \mu_S S_t dt + \sigma_S S_t dZ_t^S \tag{4}$$

where $S_t$ denotes the price at time $t$ of the reference portfolio, $\mu_S$ is the continuously compounded market rate, $\sigma_S$ is the constant volatility parameter, is a Standard Brownian Motion with the property $Cov(dZ_t^r, dZ_t^S) = \varphi dt$, $\varphi \in R$. We describe the evolution in time of mortality by a widely used mortality model supposing that the force of mortality at time $t$ for an individual aged $x + t$ at time $t$ is given by

$$d\mu_{x+t} = k^m(\gamma - \mu_{x+t})dt + \sigma^m \sqrt{\mu_{x+t}} dZ_t^m \tag{5}$$

where $\mu_{x+t}$ is the hazard rate for an individual aged $x + t$ in the year $t$, $k^m$ and $\sigma^m$ are positive constants, $\gamma$ is the long term $Z_t^m$ is a Standard Brownian motion. This model, referred as the CIR mortality model has the property that the mortality rates are continuous and remain positive. Moreover, the mortality rates does not reach zero and the drift factor ensures the mean reversion.

## 4 Numerical Results

We consider a generic subscriber entering the fund at 40 years and the outgoing age is fixed as $x + \xi = 65$. At this age begins the decumulation phase. The risk adjusted parameters of the Cox-Ingersoll-Ross process are $r_0 = 0.0273$, $k^r = 0.001$, $\sigma^r = 0.035$, $\theta = 0.0239$. The time series related to the interest rates consists of the annualized net Euribor rates covering the period from January 2003 to December 2013. Referring to the time evolution of the reference fund, we let $\mu_S = 0.03$ and $\sigma_S = 0.20$. For the correlation coefficient, a slightly negative value is adopted in line with the literature concerning the Italian stock market. With regard to the survival probabilities our data set relates to the Italian male population with age-specific death counts ranging from ages 65 over the period 1954–2008. We refer to the class of forward mortality models. These models study changes in the mortality curve for a specific age cohorts and capture dynamics of each age cohort over time for all ages greater than $x$ in a specific year $t$. In this case the mortality curves are modeled diagonally (for example see [3]). Based on our data, the Fund should provide a 2.75 % return. In monetary terms, the annual benefits are 2.5 times the contributions paid. At first glance, the investment for the member appears to be more than satisfactory. But the risk of the Fund is very high. Because of longevity, the fund is able to guarantee less than 90 per cent of the benefit promised. At this point, the benchmark level of 3 % is no longer appropriate and the fund need a more aggressive management. The fund reaches an equilibrium between the benefits promised and the contribution received only increasing by 20 % the return of the investments.

## References


1. Black, F., Scholes, M.: The pricing of option and corporate liabilities. J. Polit. Econ. 637–654 (1973)
2. Cox, J., Ingersoll, J., Ross, S.: A theory of the term structure of interest rates. Econometrica 385–408 (1985)
3. Dahl, M.: Stochastic mortality in life insurance: market reserves and mortality linked insurance contracts. Insur. Math. Econ. **35**, 113–136 (2004)
4. Denuit, M., Devolder, P., Goderniaux, A.: Securitization of longevity risk. Pricing survivor bonds with Wang transform in the Lee-Carter framework. J. Risk Insur. **74**, 87–113 (2007)
5. Denuit, M., Haberman, S., Renshav, A.: Longevity indexed life annuities. N. Am. Actuar. J. **15**, 97–111 (2011)
6. Pitacco, E.: From benefit to guarantees: looking at life insurance products in a new framework. Working paper available at: http://ssrn.com/abstract=2200310 (2013)

# Pricing and Hedging Basket Options Under Shifted Asymmetric Jump-Diffusion Processes

**Tommaso Paletta, Arturo Leccadito, and Radu Tunaru**

**Abstract** The empirical characteristics of the underlying asset prices should be taken into account for the pricing and hedging of options. In this paper, we show how to price basket options when assets follow the "shifted asymmetric jump-diffusion" process. The methodology is based on the Hermite polynomial expansion that can match exactly the first $m$ moments of the model implied-probability distribution. The resultant pricing and hedging formulae are in closed-form and similar to the Black and Scholes ones.

**Keywords** Basket options · Shifted asymmetric jump-diffusion · Hermite polynomials · Option pricing and hedging

## 1 Introduction

Basket options are contingent claims on a group of assets, commonly traded over-the-counter to hedge away exposure to correlation or contagion risks. The required multi-dimensional framework makes the pricing and hedging of these contingent claims extremely difficult and most of the existing methods sacrifice the realism of the underlying price models to circumvent these difficulties.

Recently, Borovkova et al. [1] have considered options on basket whose assets follow the geometric Brownian motion and assumed the entire basket as a single asset with "shifted log-normal" dynamics. Even though the latter model can incorporate negative skewness (which is well known to characterize equities) while still

T. Paletta (✉) · R. Tunaru
Business School, University of Kent, Canterbury, UK
e-mail: t.paletta@kent.ac.uk

R. Tunaru
e-mail: r.tunaru@kent.ac.uk

A. Leccadito
Dipartimento di Economia, Statistica e Finanza, Università della Calabria, Rende (CS), Italy
e-mail: arturo.leccadito@unical.it

C. Perna, M. Sibillo (eds.), *Mathematical and Statistical Methods for Actuarial Sciences and Finance*, DOI 10.1007/978-3-319-05014-0_38, 

retaining analytical tractability, most of the empirical characteristics of the assets are lost when the resulting dynamics for the basket is considered. In this work, we tackle this problem from a different perspective. We assume that each asset follows the "shifted asymmetric jump-diffusion" process, which can incorporate wide ranges of skewness and kurtosis values. We provide pricing and hedging formulae for basket options, by employing the Hermite polynomial expansion to match exactly the first $m$ moments of the model implied probability distribution. In the following, Sect. 2 describes the market model and Sect. 3 describes the pricing and hedging methodology and reports a numerical example in which our method is compared to the one in [1].

## 2 The Shifted Asymmetric Jump-Diffusion Model

Consider the filtered probability space[1] $(\Omega, F, (F_t)_{0\le t\le T}, P)$. Let us define, on this space, the financial market consisting of $n$ assets, $S^{(i)}$ for any $i = 1, \dots, n$, with dynamics given by

$$\begin{aligned} d\big(S_t^{(i)} - \delta_t^{(i)}\big) &= \Big(\alpha_i - \sum_{q=\{U,D\}} \lambda_{i,q}\beta_{i,q}\Big)\big(S_t^{(i)} - \delta_t^{(i)}\big)dt \\ &\quad + \big(S_t^{(i)} - \delta_t^{(i)}\big)\sum_{j=1}^{n_w} \gamma_{ij} dW_t^{(j)} \\ &\quad + \big(S_{t^-}^{(i)} - \delta_t^{(i)}\big)\sum_{q\{U,D\}} dQ_t^{(i,q)}, \quad S_0^{(i)} \text{ known}, \end{aligned} \tag{1}$$

and the bank account $dM_t = rM_t dt$ that can be used to borrow and deposit money with continuously compounded interest rate $r \ge 0$, assumed constant over time. Equation (1) describes the shifted asymmetric jump-diffusion process where $\alpha_i$ is the expected rate of return on the asset $i$, $\{W_t^{(j)}\}_{t\ge 0}$ are $n_w$ mutually independent Wiener processes, $\{Q_t^{(i,q)}\}_{t\ge 0}$ with $q = \{U, D\}$ are independent compound Poisson processes formed from some underlying Poisson processes $\{N_t^{(i,q)}\}_{t\ge 0}$ with intensity $\lambda_{i,q} \ge 0$ and $Y_j^{(i,q)}$ representing the jump amplitude of the $j$-th jump of $N_t^{(i,q)}$ for any $i = 1, \dots, n$. The index $q$ represents the source of jumps: a total of two sources were considered (indicated as $U$ and $D$). The jumps $Y_j^{(i,q)}$ for any $i = 1, \dots, n$ are independent and identically distributed random variables with probability density function $f^{(i,q)}(y) : [-1, \infty) \to [0, 1]$ and expected value under the physical measure $\beta_{i,q} = E[Y^{(i,q)}] = \int_{-1}^{\infty} y f^{(i,q)}(y)dy$. Furthermore, $\delta_t^{(i)}$ is the shift applied to $S_t^{(i)}$ and it is assumed to follow $d\delta_t^{(i)} = r\delta_t^{(i)}dt$, with $\delta_0^{(i)} \ge 0$. The cor-

[1]The content and notation in this subsection benefit from [6, Chap. 11.5].

relation among assets is defined by the constant parameters $\gamma_{ij} \in \Re$. If a solution $(\theta, \tilde{\beta}_U, \tilde{\lambda}_U, \tilde{\beta}_D, \tilde{\lambda}_D)$ of the system[2]

$$\alpha_i - \lambda_{i,U}\beta_{i,U} - \lambda_{i,D}\beta_{i,D} - r = \sum_{j=1}^{n_w} \gamma_{ij}\theta_j - \tilde{\lambda}_{i,U}\tilde{\beta}_{i,U} - \tilde{\lambda}_{i,D}\tilde{\beta}_{i,D}, \quad i = 1, \ldots, n \tag{2}$$

does exist and is selected in association with the risk-neutral pricing measure $\tilde{P}$, then, under this risk-neutral measure, the solution of (1) is

$$\frac{S_t^{(i)} - \delta_0^{(i)} e^{rt}}{S_0^{(i)} - \delta_0^{(i)}} = e^{(r - \sum_{q=\{U,D\}} \tilde{\lambda}_{i,q}\tilde{\beta}_{i,q} - \frac{1}{2}\sum_{j=1}^{n_w} \gamma_{ij}^2)t + \sum_{j=1}^{n_w} \gamma_{ij} \tilde{W}_t^{(j)}} \prod_{q=\{U,D\}} \prod_{l=1}^{N_t^{(i,q)}} (Y_l^{(i,q)} + 1). \tag{3}$$

Because the solution to (2) is, in general, not unique, we assume that one solution is selected[3] and a pricing measure $\tilde{P}$ is fixed.[4] Under the $\tilde{P}$-measure, for asset $i$-th in the basket, the intensity of the Poisson process $\{N_t^{(i,q)}\}_{t\geq 0}$ is $\tilde{\lambda}_{i,q}$ and $\tilde{\beta}_{i,q} = \tilde{E}[Y^{(i,q)}] = \int_{-1}^{+\infty} y \tilde{f}^{(i,q)}(y)dy$. An example of distributions for the two jumps can be found in Ramezani and Zeng [4] where they choose for the $U$-jump the Pareto distribution and for the $D$-jumps the Beta distribution. The two sources of jumps represent the arrival of good and bad news in the market, and cause upward and downward jumps in prices respectively.

## 3 Pricing and Hedging Methodology

Let's consider at time 0 a European put option written on a basket of the $n$ assets in the market, maturity at $T$ and strike $K^*$, then its payoff function at maturity is $(K^* - B_T^*)^+$ where $B_T^*$, the basket value, is $B_T^* = \sum_{i=1}^n a_i S_T^{(i)} = B_T + \sum_{i=1}^n a_i \delta_0^{(i)} e^{rT}$, $(a_1, \ldots, a_n)$ is the vector of basket weights, which could be positive or negative and $B$ is the "shifted basket". Equivalently, the payoff function can be written as $(K - B)^+$ where $K$, the "shifted strike price", is $K = K^* - \sum_{i=1}^n a_i \delta_0^{(i)} e^{rt}$.

The pricing method is based on the substitution of the model implied risk-neutral distribution of the basket return with the Hermite polynomial that matches exactly its first $m$ moments ($H_k(\cdot)$ denotes the $k$-th order Hermite polynomial):

$$\frac{B_T}{B_0 e^{rT}} \longrightarrow J(z) = \sum_{k=0}^{m-1} \varphi_k H_k(Z) \tag{4}$$

---

[2]The proofs of this and all other results are provided by the authors upon request.

[3]For a review on selection of pricing measures, see [2] and references within.

[4]Henceforth, $\tilde{E}$ is used to indicate the expectation operator under the risk-neutral measure $\tilde{P}$.

**Table 1** The table reports the prices of 3 basket put options written on 6 assets ($i = 1\ldots6$) which follow the shifted asymmetric jump-diffusion process with $S_0^{(i)} = i$, $\tilde{\lambda}_{i,U} = 0.4$, $\tilde{\lambda}_{i,D} = 0.04$, $Y_l^{(i,U)} + 1 \sim$ Pareto(60), $Y_l^{(i,D)} + 1 \sim$ Beta(70, 1), and $\delta_0^{(i)} = e^{-rT}$. Furthermore, $K^* = 21$, $r = 4$ % and the entries of the covariance matrix $\Sigma_{ij}$ are 0.7613, 0.8566, 0.8791, 0.6967, −0.3044, 0.4050, 0.5108, 0.4331, 0.1701, 0.9067, 0.8830, −0.6870, 0.7479, −0.5702, −0.7557 for $j = i + 1, \ldots, 6$. "MC" stands for the standard Monte Carlo method with $10^6$ simulations, "std" is its standard deviation, "BPW" is the method in [1] and "our method" is the pricing method described in this chapter

| Maturity | MC | std | BPW | Our method |
|---|---|---|---|---|
| $T = 1$ | 0.5473 | (0.0040) | 0.6078 | 0.5125 |
| $T = 2$ | 0.6382 | (0.0051) | 0.7162 | 0.6402 |
| $T = 3$ | 0.6791 | (0.0058) | 0.7465 | 0.6784 |
| RMSE | | | 0.0690 | 0.0201 |

where the $\varphi_k s$ are calculated by moment matching. Under these settings, the price of a European put basket option is given by:

$$p_0 = Ke^{-rT}\Phi(h_2\tilde{z}) - B_0\big[\varphi_0\Phi(h_2\tilde{z}) - h_2 g(\tilde{z})\big] \tag{5}$$

where $g(\tilde{z}) = \phi(\tilde{z})\sum_{k=0}^{m-2}\varphi_{k+1}H_k(\tilde{z})$, $h_2 = sgn(B_0)$, $\tilde{z}$ is the solution of $J(\tilde{z})Be^{rT} = K$, $\phi(\cdot)$ is the standard normal density function and $\Phi(\cdot)$ is the standard normal cumulative distribution function and $Z$ is a standard normal random variable. The model implied $k$-moment of $B_t$ under $\tilde{P}$, after the changing of variables $\sigma_i^2 = \sum_{j=1}^{n_w}\gamma_{ij}^2$ and $V_t^{(i)} = \sum_{j=1}^{n_w}\frac{\gamma_{ij}}{\sigma_i}\tilde{W}_t^{(j)}$, is:

$$\tilde{E}\big[B_t^k\big] = \sum_{i_1=1}^{n}\cdots\sum_{i_k=1}^{n}\left[mgf(\mathbf{e}_{i_1} + \cdots + \mathbf{e}_{i_k})\prod_{l=1}^{k}a_{i_l}\big(S_0^{(i_l)} - \delta_0^{(i_l)}\big)e^{(r+\omega_{i_l})t}\right] \tag{6}$$

where $\omega_j = -\tilde{\lambda}_{j,U}\tilde{\beta}_{j,U} - \tilde{\lambda}_{j,D}\tilde{\beta}_{j,D} - \frac{1}{2}\sigma_j^2$, $\mathbf{e}_j$ is the vector having 1 in position $j$ and zero elsewhere, the moment generation function (mgf) of $\sigma_i V_t^{(i)} + \sum_{q=\{U,D\}}\sum_{l=1}^{N_t^{(i,q)}}\log(Y_l^{(i,q)} + 1)$ is given by:

$$mgf(\mathbf{u}) = \exp\{t\mathbf{u}'\Sigma\mathbf{u}/2\}\prod_{q=\{U,D\}}\prod_{i=1}^{n}mgf_{N_t^{(i,q)}}\big(cgf_{\log(Y_{i,q}+1)}(u_i)\big) \tag{7}$$

where $\Sigma$ denotes the covariance matrix of $V = (V_t^{(1)}, \ldots, V_t^{(n)})'$, $mgf_{N_t^{(i,q)}}(u) = \exp(t\tilde{\lambda}_{i,q}(e^u - 1))$ and $cgf_{\log(Y_{i,q}+1)}(u_i)$ is the cumulant-generating function of $\log(Y_{i,q} + 1)$. For the sake of brevity, we omit the calculation of the Greek parameters that can be calculated by simple derivation of (5) and by using the methodology in [1] for the derivatives of the parameters $\varphi_j$s.

The usefulness of a pricing method can be gauged by comparing it with other established methods in the literature. For this reason, we directly benchmark our method with the one in [1] (BPW), which is a good competitor because it covers negative skewness and works well under highly multi-dimensional frameworks. Furthermore, the standard Monte Carlo methodology (see [3]) is used to obtain the "true" fair no-arbitrage price. Table 1 contains the prices for three different basket put options and also, on the last row, the root mean square error (RMSE). On this example our new method outperforms the BPW using the RMSE as a comparison criterion by about a factor of 3.5. In an extended version of this chapter, we will consider a more general comparison on a larger set of basket options and including also other underlying models such as the ones in [5].

## References

1. Borovkova, S., Permana, F., Weide, H.V.: A closed form approach to the valuation and hedging of basket and spread option. J. Deriv. **14**(4), 8–24 (2007)
2. Frittelli, M.: The minimal entropy martingale measure and the valuation problem in incomplete markets. Math. Finance **10**(1), 39–52 (2000)
3. Glasserman, P.: Monte Carlo Methods in Financial Engineering, vol. 53. Springer, Berlin (2004)
4. Ramezani, C., Zeng, Y.: Maximum likelihood estimation of the double exponential jump-diffusion process. Ann. Finance **3**(4), 487–507 (2007)
5. Rombouts, J.V.K., Stentoft, L., Violante, F.: The value of multivariate model sophistication: an application to pricing Dow Jones Industrial Average options. CORE Discussion Papers 2012003, Universitè Catholique de Louvain, CORE (2012)
6. Shreve, S.E.: Stochastic Calculus for Finance II: Continuous-Time Models. Springer Finance. Springer, Berlin (2004)

# On a Data Mining Framework for the Identification of Frequent Pattern Trends

**Marina Resta**

**Abstract** The work discusses a data mining framework that combining Self Organizing Maps (SOM) and Graphs paradigms is able to offer insights on the clusters structure of the mapping. The basics of the method rely in the use of trained SOM to define graphs from best matching units. In particular, we discuss the application to best matching units of two graphs topologies, originating the SOM-based Minimum Spanning Tree (SOM-MST), and the SOM-based Planar Maximally Filtered Graph (SOM-PMFG), respectively. We show that, working with financial time-series data, it is possible to capture the clusters structure of market assets, and to use such information for market active tradings. The discussion of results obtained working with stocks from Milan Stock Exchange concludes.

**Keywords** Self organizing maps · Minimum spanning tree · Planar maximally filtered graph · Financial time-series

## 1 Motivation

Practitioners attempting to study the behavior of financial markets make use of various instruments: quantitative finance as well as fundamental techniques, going deepest in detail of either intrinsic mathematical or budgeting features of market assets try to explain how they can affect market prices. However, there are so many factors interacting at any time that it may happen to ignore important ones, in favor of those that are considered as a kind of flavor of the day. Moreover, visually watching to financial markets, it becomes obvious that there are patterns that repeat over time: to this extent, we can claim that charts mirror the mood of the crowd, i.e. of the greatest part of the investors acting into the market, and not of the fundamental factors. Starting from this point, we are going to discuss a data mining approach that combines the analysis of financial data with visualization techniques. The beauty of the method relies on two aspects: (i) we focus on quantitative techniques that make

M. Resta (✉)
DIEC, University of Genova, via Vivaldi 5, Genova, Italy
e-mail: resta@economia.unige.it

C. Perna, M. Sibillo (eds.), *Mathematical and Statistical Methods for Actuarial Sciences and Finance*, DOI 10.1007/978-3-319-05014-0_39,

**Algorithm 1** The SOM algorithm explained

**repeat**

1 At each step $t$, present an input $\mathbf{x}(t) \in X$ and select the winner: $\nu(t) = \arg\min_i \|\mathbf{x}(t) - \mathbf{w}_i(t)\|$

2 Update the weights of the winner and its neighbors: $\Delta\mathbf{w}_i(t) = \alpha(t)\eta(\nu, i, t)(\mathbf{x}(t) - \mathbf{w}_\nu)$.

**until** the map converges.

**Algorithm 2** Gb-SOM

1 Extract from a SOM all the best matching units (BMUs)

2 For each couple of BMUs compute the correlation

3 From the correlation matrix derive the adjacency matrix and hence build the graph.

possible to group data according to their similarity in a way as objective as possible; (ii) we discuss an approach with higher visual impact; this allows final users to easily understand the results obtained and hence to use them for market active tradings. According to this rationale, what remains of the paper is organized as follows. Section 2 provides a short mathematical background. Section 3 draws the basics of an application on stocks from Milan Stock Exchange, while Sect. 4 concludes.

## 2 Mathematical Background

Self Organizing Maps [2] (SOM) assume to order a set of neurons, often arranged in a 2-D rectangular or hexagonal grid, to form a discrete topological mapping of an input space $X \subset \mathbb{R}^n$. Let us indicate by $\mathbf{w}_i \in \mathbb{R}^n$ $(i = 1, \ldots, M)$ the weight vector associated to neuron $i$; at the start of the learning, all the weights $\{\mathbf{w}_1, \mathbf{w}_2, \ldots, \mathbf{w}_M\}$ are initialized to small random numbers. Then the algorithm works as shown in the Box 1.

Here $\eta(\nu, i, t) = \exp(-\frac{\|r_\nu - r_i\|^2}{2\sigma^2})$ is the neighborhood function between neurons $r_\nu$ and $r_i$, with $\sigma$ representing the effective range of the neighborhood; while $\alpha(t)$ is the so called learning rate, that is a scalar-valued function, decreasing monotonically, and satisfying: (i) $0 < \alpha(t) < 1$; (ii) $\lim_{t\to\infty} \alpha(t) \to +\infty$; (iii) $\lim_{t\to\infty} \alpha(t) \to 0$ [2, 4].

By construction SOM is a data mining tool that preserves input topology structure, as to say: similar inputs are mapped into neighbor nodes. However by combining SOM to proper graph structures [1], it is possible to enhance such features. The pseudo-code in Box 2 explains how the procedure that we named Graph-based SOM (Gb-SOM) works.

Clearly, depending on the filtering applied in step 3 to build the adjacency matrix, different graphs can be obtained and hence different information can be retrieved.

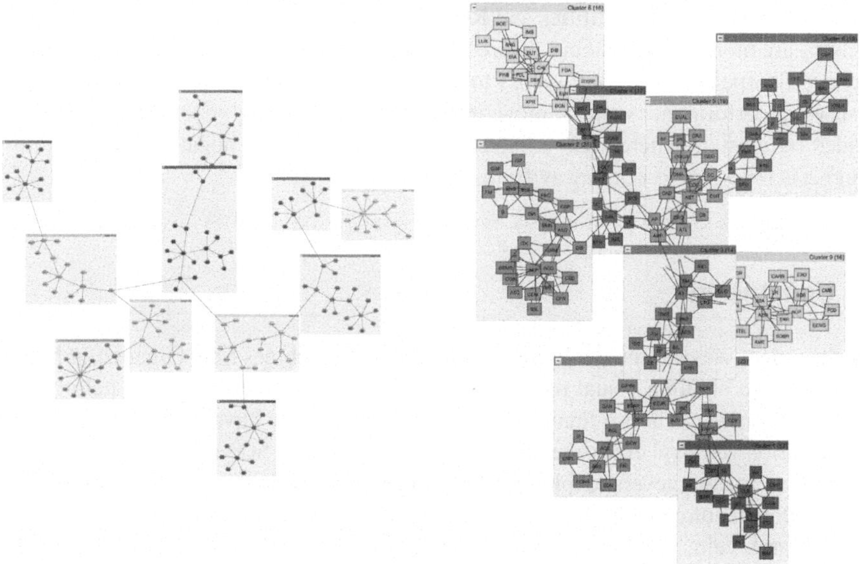

(*a*) (*b*)

**Fig. 1** From *left* to *right*: SOM-MST (**a**) and SOM-PMFG (**b**) obtained using as input matrix the log-returns of 167 traded assets in the Milan Stock Exchange

Since our aim is to extract relevant information (and not necessarily all the available information) to trade the market , we focused on two procedures: the Minimum Spanning Tree (MST) and the Planar Maximally Filtered Graph (PMFG) [5], originating the SOM-based Minimum Spanning Tree (SOM-MST), and the SOM-based Planar Maximally Filtered Graph (SOM-PMFG).

## 3 A Practical Application

While Gb-SOM has been already successfully employed to monitor countries exposure to financial crisis risk [3], we now focus on how to use it for market active tradings. To illustrate how our methodology works, we briefly detail an experiment where input data are the log-returns of 167 stocks quoted on the Milan Stock Exchange in the period: December 2011–December 2012. On those data we built for exemplification purposes both SOM-MST and SOM-PMFG. Figure 1 shows the market structure as obtained by way of SOM-MST (Fig. 1(a)) and SOM-PMFG (Fig. 1(b)): clusters are highlighted in both cases.

In both cases the market skeletonization offers visual insights on how the market organizes; in particular, whereas the SOM-MST finds 11 relevant stocks clusters, the SOM-PMFG highlights 9 groups of stocks. From the analysis of clusters composition, we find out that SOM-MST emphasizes dominating sectors (namely:

Heavy Industry, Public Utilities and Real Estates); on the contrary, in SOM-PMFG groups are more heterogeneous. Moreover, if we combine clusters composition with related Sharpe ratio values, it is easy to identify groups (and hence assets) with highest/worst performances and to move in the market accordingly. An example is provided by clusters where Heavy Industry is dominant whose negative Sharpe Ratio suggests investors to keep away from.

## 4 Conclusion

In this paper we presented a hybrid procedure that combines Self-Organizing Maps to Graphs to obtain a visual representation of financial data that can be helpful for traders in order to choose financial assets to invest on. We have given evidence that information retrieval from the obtained SOM-MST and SOM-PMFG can be performed at various levels, exploring the groups composition, as well as combining the clusters features with financial indicators. In our opinion this can be a very fruitful research vein, provided the need of deepest investigations with graphs analysis tools.

## References

1. Diestel, R.: Graph Theory, 3rd edn. Springer, Berlin (2005)
2. Kohonen, T.: Self-Organizing Maps. Springer, Berlin (2001). Third, extended edition
3. Resta, M.: The shape of crisis lessons from self organizing maps. In: Kahraman, C. (ed.) Computational Intelligence Systems in Industrial Engineering. Atlantis Computational Intelligence Systems, pp. 535–555. Atlantis Press, Paris (2012)
4. Ritter, H., Schulten, K.: Convergence properties of Kohonen's topology conserving maps: fluctuations, stability, and dimension selection. Biol. Cybern. **60**, 59–71 (1988)
5. Tumminello, M., Aste, T., Di Matteo, T., Mantegna, R.N.: A tool for filtering information in complex systems. Proc. Natl. Acad. Sci. USA **102**(30), 10421–10426 (2005)

# Risk Processes with Normal Inverse Gaussian Claims and Premiums

**Dean Teneng and Kalev Pärna**

**Abstract** We study risk processes where claims and premiums are modeled by independent normal inverse Gaussian (NIG) Lévy processes; claims by a spectrally positive NIG Lévy process. Using martingale technique, the Lundberg inequality for ruin probability is proved.

**Keywords** NIG · Risk process · Cramer-Lundberg

## 1 Introduction

Boykov [1] studies risk processes where claims and premiums are modeled by independent compound Poisson processes. In [4], the difference of premiums and claims are modeled by different Lévy processes; capitalizing on the NIG. Of recent, Stanojevic and Levajkovic [6] proposed modeling premiums with a time changed subordinated Lévy process. We implement this proposal using NIG-Lévy process since it can be represented as an inverse Gaussian time changed Brownian motion with drift [5]. Further, we model aggregate claims by a spectrally positive NIG Lévy process. Using martingale technique, we prove the Lundberg inequality for ruin probability.

## 2 Model Considerations

### *2.1 Modified Premium Process*

Generally, premiums are determinate, discrete, independent, non-negative stationary increments [2] and we consider an infinite number collected within the time period.

D. Teneng (✉) · K. Pärna
Institute of Mathematical Statistics, University of Tartu, J. Liivi 2-518, Tartu, Estonia
e-mail: dean_teneng@yahoo.com

K. Pärna
e-mail: kalev.parna@ut.ee

C. Perna, M. Sibillo (eds.), *Mathematical and Statistical Methods for Actuarial Sciences and Finance*, DOI 10.1007/978-3-319-05014-0_40,

Modified premium incorporates the effects of inflation, dividend payouts, tax, interest rate fluctuations, claims processing costs and other administrative expenses by the insurance company but claims. It can take on negative or positive values and can be represented by a function with support on the entire real line. Our proposal is to use a finite mean and finite variance NIG Lévy process; with the mean representing constant loaded premium and variance the variation in this modified premium process i.e. stochastic premiums. This is because NIG Lévy process has paths composed of an infinite number of small jumps and exhibit diffusion-like feature with a jump driven structure [3]. NIG[1] Lévy process[2] has its Laplace exponent through Lévy-Khintchine theorem [4] as follows:

$$\Psi_1(\lambda) = a\lambda + \int_{-\infty}^{+\infty} \left[e^{\lambda x} - 1 - \lambda x I_{\{|x|\leq 1\}}\right] \nu(dx) \tag{1}$$

where $a$, $\lambda$ are real constants and $\nu(dx)$ is a measure on $R \setminus \{0\}$ such that $\int_{-\infty}^{+\infty}(1 \wedge X^2)\nu(dx) < \infty$.

## 2.2 Claims

Claims also are generally independent,[3] indeterminate, stationary, non-negative increments with an infinite number collected within considered finite time interval [2]. We model these with a spectrally positive NIG Lévy process i.e. an NIG Lévy process with no negative jumps and chosen to have finite mean, finite variance and support on the positive real line (see Fig. 1). It is not the negative of a subordinator.[4] Generally, if $X$ is spectrally positive, then $-X$ is spectrally negative. Spectrally negative Lévy processes are well studied in the literature. NIG spectrally negative has its Laplace exponent through Lévy-Khintchine theorem as follows:

$$\Psi_2(\lambda) = -a_1\lambda + \int_{-\infty}^{0} \left[e^{\lambda y} - 1 - \lambda y I_{\{y > -1\}}\right] \nu_1(dy) \tag{2}$$

[1] A random variable $X$ is NIG distributed, (denoted NIG$(\alpha, \beta, \delta, \mu)$) if its probability density function is given by $f_{\text{NIG}}(x; \alpha, \beta, \delta, \mu) = \frac{\alpha\delta}{\pi} e^{\delta\sqrt{\alpha^2-\beta^2}+\beta(x-\mu)} \frac{K_1(\alpha\sqrt{\delta^2+(x-\mu)^2})}{\sqrt{\delta^2-(x-\mu)^2}}$ where $K_\lambda(x) = \int_0^\infty (u^{\lambda-1} e^{\frac{-x}{2}(u+u^{-1})})du$ with $\delta > 0$ scaling parameter, $\alpha > 0$ shape parameter, $\beta$ with $0 \leq |\beta| \leq \alpha$ skewness parameter and $\mu \in \Re$ location parameter. The mean and variance are given by $\mu + \frac{\delta\beta}{\sqrt{\alpha^2-\beta^2}}$ and $\delta\frac{\alpha^2}{[\sqrt{\alpha^2-\beta^2}]^3}$ respectively. It has a simple Laplace exponent $\Psi(\lambda) = -\mu\lambda + \delta(\sqrt{\alpha^2-\beta^2} - \sqrt{\alpha^2-(\beta-\lambda)^2})$, $|\beta - \lambda| < \alpha$.

[2] Let $(\Omega, F, (F(t))_{t\geq 0}, P)$ be a filtered probability space. An adapted *cádlág* $\Re$-valued process $X = \{X(t)\}_{t\geq 0}$ with $X(0) = 0$ is NIG Lévy process if $X(t)$ has independent stationary [3] increments distributed as NIG$(\alpha, \beta, \delta, \mu)$.

[3] We leave out cases of disasters and serial accidents where claims can be correlated.

[4] A subordinator is a strictly non-decreasing Lévy process.

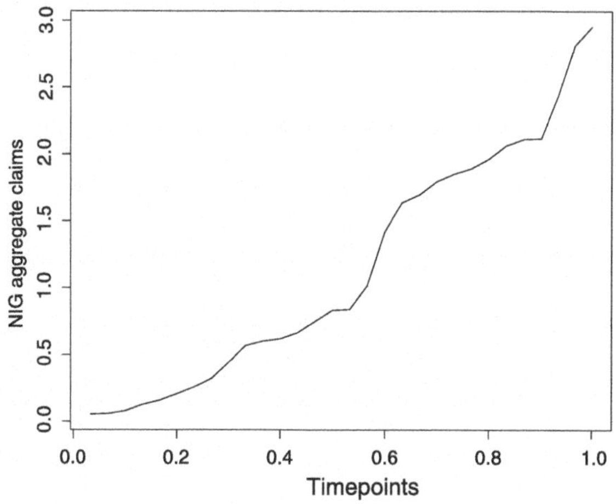


**Fig. 1** Spectrally positive NIG(50, −10, 1, 0) process depicting aggregate claims

where $a_1$ is a real constant representing drift, $\lambda$ also real and $\nu_1(dy)$ is a measure on $R \setminus \{0\}$ such that $\int_{-\infty}^{\infty}(1 \wedge Y^2)\nu_1(dy) < \infty$.

# 3 Proposed Model

We propose a Cramer-Lundberg risk model with modified stochastic premiums and stochastic claims both modeled by different NIG-Lévy processes. The risk process $U(t), t \geq 0$ is defined as

$$U(t) = u + X_1(t) - X_2(t) \tag{3}$$

where $u$ is the initial capital, $X_1(t) \sim \text{NIG}(\alpha, \beta, \delta_1 t, \mu_1 t)$ the modified premium process and $X_2(t) \sim \text{NIG}(\alpha, \beta, \delta_2 t, \mu_2 t)$ the aggregate claims process. The risk process $U(t) - u$ is distributed as $\text{NIG}(\alpha, \beta, (\delta_1 - \delta_2)t, (\mu_1 - \mu_2)t)$ with its Laplace exponent through Lévy-Khintchine theorem:

$$\Psi_T(\lambda) = \Psi_1(\lambda) + \Psi_2(\lambda) = (a - a_1)\lambda + \int_0^{+\infty} \left[e^{\lambda x} - 1 - \lambda x I_{\{|x| \leq 1\}}\right]\nu(dx) \tag{4}$$

taking in to account the chosen approximation $\int_{-\infty}^{0}[e^{\lambda x} - 1 - \lambda x I_{\{|x| \leq 1\}}]\nu(dx) \approx \int_{-\infty}^{0}[e^{\lambda y} - 1 - \lambda y I_{\{y > -1\}}]\nu_1(dy)$. This approximation basically means the company has controls in such a way that most claims can be settled, but ultimate ruin or profitability is determined by (4). $\Psi_T(\lambda)$ of (4) has both a constant part $(a - a_1)$ and a stochastic part $\int_0^{+\infty}[e^{\lambda x} - 1 - \lambda x I_{\{|x| \leq 1\}}]\nu(dx)$ representing in a sense the modified risk process. Therefore, net profitability condition simply translates to $E(X(1)) > E(Y(1))$, where $X(1)$ represents modified premium and $Y(1)$ aggregate claims.[5]

[5]Making use of infinite divisibility property of Lévy processes.

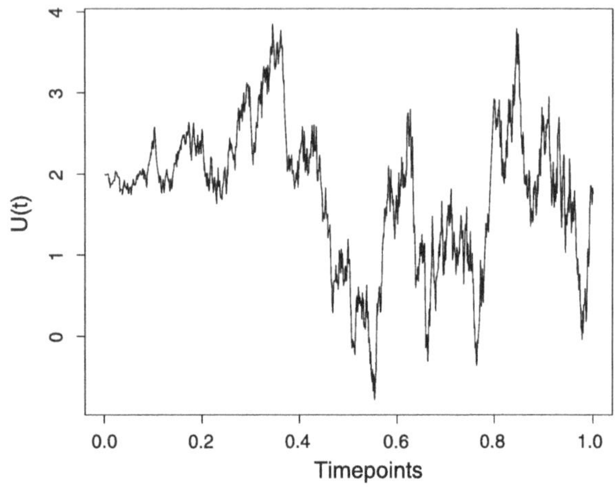


**Fig. 2** Risk process which is the difference of NIG(50, −10, 1, 0) and spectrally positive NIG(50, −10, 2, 0)

Considering the Martingale[6] approach to ruin probability, if we can find a value $r = R$ in the domain of the definition of $\Psi(\lambda)$ such that $\Psi(\lambda) = 0$ and $\tau < \infty$, then we could simply write

$$\psi(u) = E_Q\big[e^{RU(\tau)}\big]e^{-Ru}, \quad u \geq 0. \tag{7}$$

We calculate such an $R = \frac{2(d\sqrt{\alpha^2-\beta^2}+\beta)}{d^2+1}$ where $d = \frac{\mu_1-\mu_2}{\delta_2-\delta_1}$, similar to [4] where their $c = \mu_1 - \mu_2$ and $\delta = \delta_2 - \delta_1$; employing similar analysis. Simulated graph (Fig. 2) demonstrates how the proposed model looks like.

**Acknowledgements** Research financed by Estonian Science Foundation grant 8802 and Estonian Doctoral School in Mathematics and Statistics.

## References

1. Boykov, A.V.: Cramér-Lundberg model with stochastic premiums. Teor. Veroâtn. Ee Primen. **47**, 549–553 (2002)

---

[6]If $X$ is a Levy process with Laplace exponent $\Psi(\lambda)$, then

$$M_t^r = \frac{e^{-rX(t)}}{e^{t\Psi(r)}} \tag{5}$$

with $r$ a real number is a local martingale. Ruin probability is defined as $\psi(u) = P\{\tau < \infty\}, u \geq 0$ and $\tau = inf\{t > 0 : U(t) < 0\}$. Using Esscher transform [4, 5] to change the martingale $M_t^r$ from $P$ to a locally equivalent measure $Q$ for some $r$ in the domain of the Laplace exponent, we can write

$$\psi(u) = E_P[I_{\{\tau<\infty\}}] = E_Q\left[\frac{I_{\{\tau<\infty\}}}{M_\tau^r}\right] = E_Q\big[e^{r(U(\tau)-u)+\tau\Psi(r)}I_{\{\tau<\infty\}}\big]. \tag{6}$$

Hence, if we can find a value $r = R$ in the domain of the definition of $\Psi(r)$ such that $\Psi(r) = 0$ and $\tau < \infty$, then we could simply write $\psi(u) = E_Q[e^{RU(\tau)}]e^{-Ru}, u \geq 0$.

2. Dufresne, F., Gerber, H.U., Shiu, E.S.W.: Risk theory with gamma process. ASTIN Bull. **21**, 177–192 (1997)
3. Godin, F., Mayoral, S., Morales, M.: Contingent claim pricing using a normal inverse Gaussian probability distortion operator. J. Risk Insur. **79**, 841–866 (2012)
4. Morales, M., Schoutens, W.: A risk model driven by Lévy processes. Appl. Stoch. Models Bus. Ind. **19**, 147–162 (2003)
5. Schoutens, W.: Lévy Processes in Finance. Wiley, New York (2003)
6. Stanojevic, J., Levajkovic, T.: On the Cramér-Lundberg model with stochastic premia and the Panjers recursion. In: Cuculescu, I., Jaric, J., Gavruta, P., Golet, I., Cadariu, L. (eds.) Proceedings of the 13th International Conference on Mathematics and Its Applications, vol. 42, pp. 295–302. Editura Politehnica, Bucharest (2013)

# A Portfolio Model for the Risk Management in Public Pension

**Tadashi Uratani**

**Abstract** The financial viability of government pension plan implies that the reserve of pension fund should be positive in the demographic and economical environment change, under the condition that the income replacement ratio is more the given level. Assuming the market asset and the income for pension follows Ito processes and the population are modeled by cohort, we apply the martingale method of the optimal consumption and investment theory to guarantee the pension fund positivity.

**Keywords** Pension · Risk management · Martingale

## 1 The Model

Let $p(t, y)$ denote the numbers of policyholders of the age $y$ at $t$ and $\omega_1$ the starting age of paying premium, $\omega_2$ the end age and starting of receiving benefit $\omega_3$ the end age of beneficiary. The total number of contributors satisfies: $\xi_t^1 = \int_{\omega_1}^{\omega_2} p(t, y)dy$. The total number of beneficiaries: $\xi_t^2 = \int_{\omega_2}^{\omega_3} p(t, y)dy$. The balance of total premium and benefit is assumed to be based on the average wage. Let $H_t$ denote the average wage at $t$ and $a_t$ be the rate of premium. Let $u_t$ be the total premium amount at $t$; $u_t := a_t H_t \xi_t^1$. Let $b_t$ be the benefit ratio to the average wage and $s_t$ be the total benefit amount; $s_t := b_t H_t \xi_t^2$. We assume that $a_t, b_t$ is predictable process and it satisfies self-finance strategies, which are in the condition of $0 < a_t, b_t < 1$. The balance of premium and benefit $q_t(a, b)H_t$ is defined as:

$$u_t - s_t = \left(a_t \xi_t^1 - b_t \xi_t^2\right) H_t =: q_t(a, b) H_t.$$

The pension portfolio consists of three assets; Market asset price satisfies the following Itô process:

$$dA_t / A_t = \mu_r(t)dt + \sigma_r(t)dW_t^r =: dr_t.$$

T. Uratani (✉)
Faculty of Engineering and Science, Hosei University, Tokyo, Japan
e-mail: uratani@hosei.ac.jp

C. Perna, M. Sibillo (eds.), *Mathematical and Statistical Methods for Actuarial Sciences and Finance*, DOI 10.1007/978-3-319-05014-0_41,

Human capital price(wage) process satisfies:

$$dH_t/H_t = \mu_x(t)dt + \sigma_x(t)dW_t^x =: dx_t.$$

Let $r$ the risk free rate and the money market asset be $e^{rt}$. Portfolio strategies of the pension fund are denoted as $\pi_t := (\phi_t, a_t, b_t, \beta_t)$; Let $\phi_t$ denote the investment amount to market asset, $(a_t, b_t)$ denote the strategies for the human capital which means the policy of pension. Let $\beta_t > 0$ denote the government subsidy to pension fund at $t$ and $R_t$ denote the value of pension fund at $t$. The portfolio value satisfies at $t$:

$$R_t = \phi_t A_t + q(a,b)H_t. \tag{1}$$

From the predictability of strategies $(\phi_t, \beta_t)$,the dynamics of fund becomes;

$$dR_t = \phi_t dA_t + d\big(q(a,b)H_t\big) + \beta_t dt, \quad R_0 = \bar{R}. \tag{2}$$

The population dynamics is assumed to be non stochastic but not satisfies the self financing condition then the dynamics of the balance of premium and benefit of pension is as follows,

$$d\big(q(a,b)H_t\big) = q(a,b)dH_t + dq(a,b)H_t,$$

where $dq(a,b) := a_t d\xi_t^1 - b_t d\xi_t^2$. By using (1) we obtain:

$$dR_t = R_t dr_t + q(a,b)H_t(dx_t - dr_t) + dq(a,b)H_t + \beta_t dt.$$

From the PDE of McKendrick-von Foerster in [2], $\frac{\partial p(t,y)}{\partial t} = -\frac{\partial p(t,y)}{\partial y} - \mu(y)p(t,y)$, where $\mu(y)$ is decreasing speed of pensionaries of the age $y$. We use the method of characteristics in PDE which equals to use the cohort model of population, as $t = k + y$ and $v(k,y) := p(t,y)$.

$$dv(k,y) = -\mu(y)v(k,y)dy, \quad v(k,y) = v(k,0)\exp\left(\int_0^y \mu(s)ds\right).$$

Thus the change of pension balance satisfies

$$dq_t(a,b)H_t = -\left(a_{k+y}\int_{\omega_1}^{\omega_2^-} \mu(y)v(k,y)dy + b_{k+y}\int_{\omega_2}^{\omega_3} \mu(y)v(k,y)dy\right)H_t, \tag{3}$$

where $v(k,\omega_1)$ is the new entry numbers of contributors and $v(k,\omega_2)$ is the new entry umbers of beneficiaries.

## 2 The Optimal Pension Strategies for Cohorts

For the time horizon from 0 to $T$ there are $k$ cohort of $0 \le k \le T_l := T - (\omega_3 - \omega_1)$ who are their all contributions and benefits are within the planning period. These cohorts are new comers to the pension. Let $c(k)$ be the net present value for $k$th cohort,

$$c(k) := -\int_{\omega_1}^{\omega_2^-} a_{y+k}H_{y+k}^* v(k,y)dy + \int_{\omega_2}^{\omega_3} b_{y+k}H_{y+k}^* v(k,y)dy. \tag{4}$$

For existing pensionaries ($-\omega_3 < k < 0$), their premium and benefit was decided as $a_c$ and $b_c$ for the past: $y + k < 0$; They will follow the new premium and benefit from now $y + k \geq 0$. The net present value $c_e(k)$ should be positive;

$$c_e(k) = -\int_{\omega_1}^{\omega_2^-} (a_c \mathbf{1}_{\{y+k<0\}} + a_{y+k}\mathbf{1}_{\{y+k\geq 0\}}) H^*_{y+k} v(k,y) dy$$
$$+ \int_{\omega_2}^{\omega_3} (b_c \mathbf{1}_{\{y+k<0\}} + b_{y+k}\mathbf{1}_{\{y+k\geq 0\}}) H^*_{y+k} v(k,y) dy > 0. \quad (5)$$

For the future cohort ($T_l < k < T$) whose benefit will not finished before $T$, their net present value $c_p(k)$ should be positive;

$$c_p(k) = -\int_{\omega_1}^{\omega_2^-} (a_{k+y} \mathbf{1}_{\{y+k<T\}} + \tilde{a}_{y+k}\mathbf{1}_{\{y+k\geq T\}}) H^*_{y+k} p(y+k,y) dy$$
$$+ \int_{\omega_2}^{\omega_3} (b_{y+k} \mathbf{1}_{\{y+k<T\}} + \tilde{b}_{y+k}\mathbf{1}_{\{y+k\geq T\}}) H^*_{y+k} p(y+k,y) dy > 0. \quad (6)$$

The objective function is to maximize the utility function of the new pension participant who are the cohort of $0 \leq k \leq T_l$, where $U_1(\cdot)$ is a utility function for present value of pension and $U_2(\cdot)$ is the utility of fund value at $T$:

$$\max_{\pi_t} E\left[\int_0^{T_l} U_1\big(c(k)\big)dk + U_2(R_T)\right].$$

Beside constraints (5) and (6), we impose the following constraints seen in [3]; (1) No default of pension fund, which should satisfies the following; $R_t > 0$ $\forall t \in [0,T]$, (2) Government subsidy $\gamma_t$ should be within the limitation; $E^Q[\int_t^T e^{-rs}\beta_s ds|\mathcal{F}_t] \leq \gamma_t$.

## 3 Martingale Method for the Risk Management

The risk management of pension should be no default which implies that $R_t > 0$ for all $t \in (0,T)$. It can be treated by the martingale method of optimal investment and consumption problem of Dana-Jeanblanc [1]. Let $e^{rt}$ a numeraire from (1) and (2)

$$d\big(R_t/e^{rt}\big) = \phi_t d\big(A_t/e^{rt}\big) + q(a,b)d\big(H_t/e^{rt}\big) + dq(a,b)H_t e^{-rt} + \beta_t e^{-rt}dt.$$

Let denote $R^*_t := R_t/e^{rt}$, $A^*_t = A_t/e^{rt}$, $H^*_t = H_t/e^{rt}$, then

$$dR^*_t = \phi_t dA^*_t + q(a,b)dH^*_t + dq(a,b)H^*_t + \beta_t e^{-rt}dt.$$

The reserve fund at $T$ becomes as follows:

$$R^*_T = R_0 + \int_0^T \phi_t dA^*_t + \int_0^T q(a,b)dH^*_t + \int_0^T H^*_t dq(a,b) + \int_0^T \beta_t e^{-rt}dt,$$

where $H^*_t$ and $A^*_t$ are martingales under the risk neutral measure $Q$. The admissible strategies satisfying the constraint $R_t > 0$. The pension problem due to aging with low fertility satisfies generally the condition $dq(a,b) = a_t d\xi_t^1 - b_t d\xi_t^2 < 0$. It, however, for the cohort that the positive net present value of pension implies the condition as follows.

We assume that the government subsidy to the pension fund should satisfies $\beta_t \leq -dq(a,b)$.

$$R_t^* - \int_0^t H_s^* dq(a,b) - \int_0^t \beta_t e^{-rs} ds = R_0 + \int_0^t \phi_s dA_s^* + \int_0^t q(a,b) dH_s^* =: M_t$$

$M_t$ is a positive $Q$ Martingale

$$\begin{aligned} R_t^* &= M_t + \int_0^t H_s^* dq(a,b) + \int_0^t \beta_t e^{-rs} ds \\ R_t^* &= E^Q\left[ R_T^* - \int_t^T H_t^* dq(a,b) - \int_t^T \beta_t e^{-rs} ds | \mathcal{F}_t \right]. \end{aligned} \tag{7}$$

The necessary condition of $R_t > 0$ is $\beta_t + dq(a,b) < 0$ and $R_T^* > 0$ then

$$E^Q\left[ R_T^* - \int_0^T H_s^* dq(a,b) \right] \leq R_0 + \gamma_0.$$

## References

1. Dana, R.-A., Jeanblanc, M.: Financial Markets in Continuous Time. Springer Finance, pp. 137–144 (2007)
2. Kot, M.: Elements of Mathematical Ecology. Cambridge University Press, Cambridge (2001)
3. Ministry of Health, Labor and Welfare: The Actuarial Valuation of Employees Pension Insurance and the National Pension, Tokyo, Japan (2009, 2004)

# Black Scholes Option Sensitivity Using High Order Greeks

**Yves Rakotondratsimba**

**Abstract** Option high order sensitivities have been presented by Carr P. as *Greeks for geeks*, though other authors have analyzed and insisted on the need to go beyond to the Delta-Gamma approximation usually considered in the practice of risk management. Actually in the stress-testing framework, as is required under Basel 3 bank regulation, adding high order Greeks may contribute to a good prediction of the option PL under extreme shocks. We revisit the Black-Scholes high order Greek parameters by providing their explicit formulas and proofs, which are expected to be more accessible for many readers. Limit of the use of these sensitivities are also analyzed here. Actually our main contribution in this work is on the introduction of an unified sensitivity approach with the ones used for other classes of assets as interest rates and commodities. This may be useful in the Credit Adjustment Valuation computation and hedging, where all aspects of risk (equity, interest rate, credit, commodities, ...) need to be simultaneously considered.

**Keywords** Black-Scholes · Option · Sensitivities · P&L

## 1 Motivation

The post 2007–2008 financial crisis led the quantitative finance community to be confronted with various and increasing challenges. Indeed, the inheritance from the past remains with a lot of main issues still not completely elucidated. On the other hand, the present time requires us to be face with new market practices and regulation changes (as Basel 3 and Solvency 2). These latter call for an exploration of new approaches, though the past non-perfect tools and ideas are still considered and re-used. For example, the Black-Scholes option pricing introduced in 1973 is seen and recognized as not suitable to be used in the practice reality, however the con-

Y. Rakotondratsimba (✉)
ECE Paris Graduate School of Engineering, 37 quai de Grenelle CS71520, 75 725 Paris 15, France
e-mail: rakotond@ece.fr

Y. Rakotondratsimba
ESLSCA Business School, 1 rue Bougainville, 75007 Paris, France

C. Perna, M. Sibillo (eds.), *Mathematical and Statistical Methods for Actuarial Sciences and Finance*, DOI 10.1007/978-3-319-05014-0_42,

cept remains to be fundamental both in practice and theory. As noted in [1] *despite its age, the Black-Merton-Scholes (BMS) model is the lingua franca of option pricing. Greeks in BSM model continues to enjoy multiple applications as in hedging, market risk measurement, profit and loss attribution, model risk assessment, optimal contract design and implied parameter estimation.*

Nowadays the framework of banking new regulation Basle 3, as in the determination of Credit Valuation Adjustment (CVA) and in counterparty risks management, leads us naturally [10] to consider again the pricing and hedging of a European option on various underlying assets (as equities, bonds, swaps, ...). Facing with the associated complexity, one way very often considered by people is to come down to the Black-Scholes framework [3]. The CVA management itself leads to take under consideration the option sensitivity with respect to the joint effects of various risk-drivers as equity, interest rate, credit, commodities and so on. The sensitivities are also in the heart of fixing the capital requirement in the framework of Insurance Solvency2.[1]

Among the main ideas in risk management, always in use, is to see the future P&L of a portfolio position as a (generally nonlinear) function of one or many risk driver(s). Either for the prediction or for the position management and hedging, it may be suitable to substitute the involved function by a more simplified expression. As in the case of stress tests, constraints imposed by computation speed as well as database structure lead analysts to use approximations for the portfolio possible values. Indeed these last appear to be the result of a large variety of scenarios of asset price changes due to shocks with different sizes. One approach in direction of this simplification is to use a linear approximation, but it is also standard to resort to at least a second order polynomial function of the risk driver(s) as the case with the famous Delta-Gamma approximation (see for example [5]). However, mathematically speaking, a best fit of the P&L position function[2] would be realized by using a high order polynomial function.

The consideration of option Black-Scholes high order Greeks is not a new topic, since it has been studied by various authors as in [1, 4] and [2]. However facing with the new challenges coming from the Financial market and Insurance, as mentioned above, where extreme shocks and joint effects of risks are needed to be taken into account, then it appears that revisiting (and possibly enhancing) these high order sensitivities would be interesting and useful. To perform such a task is our intention in the present paper.

## 2 Our Contribution

As alluded above we will focus on the case of any vanilla European option whose the underlying asset is assumed to follow a log-normal process. Though this last is

[1]Further informations are available from the EIOPA web-site: https://eiopa.europa.eu/activities/insurance/solvency-ii/index.html.

[2]Which is a highly nonlinear function of the risk driver(s) when the considered position contains derivatives.

not necessarily or directly the suitable model to use in the market reality, it provides a benchmark approach which may be useful to treat some complex situations.[3]

1. Under the log-normal process, there is a one-to-one mapping between the underlying asset relative change and the corresponding shock responsible. By the term *shock* we mean a realization of some standard Gaussian random variable. Numerical results will be provided in order to better visualize this correspondence between the asset relative change and its associated shock. At this stage, it has a full sense to consider the option sensitivity with respect to the shock risk driver rather than to the asset relative change itself, which is commonly used by various authors as in [1, 4] and [8]. Such a new direction, based on the direct exploration of the underlying risk factor source, is in line with the approach we have used in the framework of interest rate in [9] and [6]. The point in this unified approach is on the opening the way of taking into account the joint effects of various risk factors. However this last aspect is not analyzed in-depth here, as we focus only on the plain Black-Scholes log-normal model.
2. Among the main differences between the present work and those performed in [1, 4] and [2] is that our sensitivities take into account the passage of time. This is in accordance with the collective intuition that the sensitivities values should differ with respect to the horizon considered. The classical Black-Scholes Delta and Gamma do not account for such a fact, and consequently this contributes to the loss of accuracy in the option P&L approximation when using these Greek parameters. Of course people make also use of the Theta parameter to take into account the passage of time, but this is not sufficient enough. The analogue of Theta, in our present approach, is a zero-order sensitivity whose the explicit expression is displayed in the text. When no shock does affect the underlying asset, then the option change is exactly reduced to this zero-order sensitivity.
3. High-order option sensitivities (with respect to the shock driver) are derived in the present work. They have a little bit resemblance with the BS high order Greeks introduced by Carr P. in [1], but the difference is that here we do not have to calculate values of Hermite polynomial functions at the points usually referred as $d_1$ and $d_2$ in the Black-Scholes pricing framework. Some recursion formulas are provided such that the reader does not need any knowledge and notions about Hermite polynomial functions to perform the calculations.
4. Once all of these sensitivities are introduced, then we state that the option P&L at a given horizon can be approximated by a polynomial function having these various high order sensitivities as coefficients and the shock as underlying variable. The main point here is that there is no a priori limitation on the size of the shock under consideration. Moreover the approximation is valid until any time-horizon strictly less than the option maturity. The classical BS Delta yields directly the proportion of underlying asset to hold or to sell in order to partially hedge the associated option. With our high order polynomial function, whose the underlying variable is the risk driver, the first order sensitivity has not the same

[3] As for instance in CVA sensitivities calculations.

interpretation. In the hedging perspective, as seen in [6]. what matters is just the compensations between the various sensitivities of the same order and for the whole position (to hedge and the hedging instruments).

5. Beside our introduction of the analytical formulas for the option high order Greeks and their implementation, three relevant issues arise: (a) Do these sensitivity parameters really contribute to improve the option P&L prediction? (b) What is the appropriate order of the polynomial function realizing a better approximation of the future option P&L? (c) What is the range of asset relative changes for which the polynomial approximation makes a full sense? Our paper also addresses to these questions by providing both theoretical analysis and numerical approaches. For example, to the first question, in contrast with [4] and [7], we find that the Taylor-Lagrange approach may provide a definitive practical implication compared with the Taylor series ones. Our answers to the next two questions rely on possible pointwise estimates for the remainder term under the user views on shock.

## References

1. Carr, P.: Deriving derivatives of derivative securities (Greeks for Geeks). J. Comput. Finance (2001). http://www.math.nyu.edu/research/carrp/papers/pdf/DDDJCFpub.pdf
2. Ederington, L., Guan, W.: Higher order Greeks (2004). http://facultystaff.ou.edu/E/Louis.H.Ederington-1/papers/Greeks_paper_Nov04.pdf
3. Elhouerkhaoui, Y.: Trading CVA: a new development in correlation modelling. Citigroup Global Markets Limited (2010). http://www.mth.kcl.ac.uk/finmath/presentations/Elouerkhaoui2011.pdf
4. Estrella, A.: Taylor, Black and Scholes: series and risk management pitfalls. Research Paper #9501, Federal Reserve Bank of New York (1995). http://www.newyorkfed.org/research/staff_reports/research_papers/9501.pdf
5. Hull, J.: Options, Futures and Other Derivatives, 8th edn. Prentice Hall, New York (2011)
6. Jaffal, H., Rakotondratsimba, Y., Yassine, A.: Hedging with a portfolio of interest rate swaps. Commun. Math. Finance **2**(1) (2013). http://www.scienpress.com/journal_focus.asp?main_id=70&Sub_id=IV&Issue=629
7. Pantz, J.: PnL, prediction under extreme scenarios (2013). http://papers.ssrn.com/sol3/papers.cfm?abstract_id=2281873
8. Rakotondratsimba, Y.: Modified delta-gamma approximation (2009). http://papers.ssrn.com/sol3/papers.cfm?abstract_id=1395969
9. Rakotondratsimba, Y.: Interest rate sensitivities under the Vasicek and Cox-Ingersoll-Ross models (2011). http://papers.ssrn.com/sol3/papers.cfm?abstract_id=1977902
10. Stein, H., Lee, K.: Counterparty valuation adjustments credit risks frontiers: subprime crisis, pricing and hedging. CVA, MBS, Rating: Tomasz Bielecki, Damiano Brigo and Frederic Patras (eds.) (2010). http://papers.ssrn.com/sol3/papers.cfm?abstract_id=1463042

The manufacturer's authorised representative in the EU is Springer Nature Customer Service Centre GmbH, Europaplatz 3, 69115 Heidelberg, Germany. If you have any concerns regarding our products, please contact ProductSafety@springernature.com

Printed and bound by CPI Group (UK) Ltd, Croydon, CR0 4YY
30/09/2026
02225567-0008